ASHEVILLE-BUNCOMBE TECHNICAL INSTITUTE
NORTH
STATE BOARD
DEPT. OF COMMUNITY COLLEGES
LIBRARIES

DISCARDED

DEC - 6 2024

*International Series of Monographs in*
CIVIL ENGINEERING
*Chief Executive Editor:* D. J. Silverleaf
*Executive Editor:* J. L. Raikes

# Concrete in Highway Engineering

*Frontispiece:* Twin three-lane carriageways on the M.1 a few miles north of London.

# Concrete in Highway Engineering

BY

D. RAYMOND SHARP

MBE, DSc, DipTP(Lond), FICE, FIMunE, FIStructE, FInstHE

**PERGAMON PRESS**
*Oxford · New York · Toronto*
*Sydney · Braunschweig*

PERGAMON PRESS LTD.
Headington Hill Hall, Oxford
PERGAMON PRESS INC.
Maxwell House, Fairview Park, Elmsford, New York 10523
PERGAMON OF CANADA LTD.
207 Queen's Quay West, Toronto 1
PERGAMON PRESS (AUST.) PTY. LTD.
19a Boundary Street, Rushcutters Bay, N.S.W. 2011, Australia
VIEWEG & SOHN GMBH
Burgplatz 1, Braunschweig

Copyright © 1970 D. R. Sharp

*All Rights Reserved. No part of this publication may be reproduced, stored in a retrieval system, or transmitted, in any form or by any means, electronic, mechanical, photocopying, recording or otherwise, without the prior permission of Pergamon Press Ltd.*

First edition 1970

Library of Congress Catalog Card No. 77-118319

*Printed in Great Britain by A. Wheaton & Co., Exeter*

**08 015845 5**

# Contents

|  |  |  |
|---|---|---|
| | PREFACE | vii |
| 1 | **Concrete as a Material** | 1 |
| | *Cement; Aggregates; Water; Concrete Mixes; Hot and Cold Weather Concreting; Curing Concrete* | |
| 2 | **Pavement Design** | 21 |
| | *Types of Pavement; Principles of Design; Traffic Loading; Design of Flexible Pavements; Design of Concrete Pavements* | |
| 3 | **Subgrade Soils, Sub-bases and Drainage** | 38 |
| | *Moisture Control and Drainage; Control of Surface and Subsoil Water; Frost-susceptible Soils; Layout for Subsoil Drainage; Layout for Surface Water Drainage* | |
| 4 | **Concrete Roads** | 45 |
| | *History; The Economics of Concrete Roads; Construction of Concrete Roads; Small-scale Projects— Semi-manual Methods; Medium-sized Projects; Large-scale Projects; Slip Form Paver* | |
| 5 | **Cement Stabilised Bases** | 82 |
| | *Introduction; Soil-cement; Lean Concrete; Cement Bound Granular Material; Treatment at Joints; Surfacing with Bituminous Materials; Compression Bumps* | |
| 6 | **Prestressed Concrete Roads** | 105 |
| | *Introduction; Advantages and Disadvantages of Prestressing; Types of Prestressed Roads; Individual Slab Type; Continuous Type; Longitudinal Prestress Requirements; Transverse Prestress; Costs* | |
| 7 | **Maintenance and Repair Techniques** | 117 |
| | *Introduction; Cement Stabilised Bases; Concrete Roads* | |

| | | |
|---|---|---|
| **8** | **The Appearance and Surface Characteristics of Concrete**<br>*Introduction; Road Surfaces; Structural Concrete* | 132 |
| **9** | **Construction in Extreme Conditions—Hot and Cold Weather**<br>*Hot Weather Concreting; Cold Weather Concreting; Cemented Materials in Hot and Cold Weather* | 151 |
| | INDEX | 161 |
| | OTHER TITLES IN THE SERIES IN CIVIL ENGINEERING | 173 |

# Preface

THERE is a shortage of books dealing with certain aspects of the design and construction side of highway work and it is hoped that this volume will fulfil a current need in its field. It is primarily intended for engineers and others who are faced with the prospect of designing or constructing cement-using roads and who already have a knowledge of highway engineering, but it should also prove helpful to architects concerned with concrete finishes and roadwork. It is anticipated that it will also be of value to students, in particular those taking post-graduate courses at universities and technical colleges.

Concrete is in ever-widening use and the highway engineering field is no exception. In addition to the more obvious use of cemented materials and concrete in road bases and slabs, concrete is now very extensively used for other highway structures, notably bridges. For this reason, some discussion is included on the appearance of concrete. Insufficient attention is still paid to producing good results on the surface of concrete when it is exposed to the natural elements and to view. This is unfortunate because once concrete has hardened the remedial action that can be taken to remove defects from the surface and improve the appearance is limited. There is a very great gulf between good and mediocre work and Chapter 8 indicates the scope of finishes at present available and how they can be best obtained. Architects are often preoccupied with this aspect of concrete work, but they are often responsible too for specifying and constructing road work and the first part of Chapter 4 and Chapter 5 may be of value to them.

Chapter 1 on Concrete as a Material is included chiefly for students, but the practising engineer and architect may find it of value as a refresher. Prestressed concrete roads are included in a separate chapter, not because the present extent of their use warrants this, but because Chapter 4 on Concrete Roads is already very comprehensive and rather long. The subjects of maintenance and repair have been given a chapter in their own right because this is a field which is much neglected by most highway engineers. The developments in the repair of concrete roads, in particular, which have taken place in the last decade now mean that repairs to this type of pavement are a practical and economic proposition. Retexturing to improve skid resistance is included under this heading.

This book is not designed to provide all the information necessary in the construction of cement-using roads, although the large number of photographs should be particularly helpful to those with limited experience. The emphasis is rather on information not readily available elsewhere and for this reason it contains little on standard techniques such as concrete mix design. A chapter on Construction in Extreme Conditions which has been included should be of particular assistance to those overseas who have to construct concrete work in conditions of heat and cold unknown in Britain.

The contents of this book have evolved as a result of the author's work in the roads field with the Cement & Concrete Association over the last fifteen years. He will always be grateful to this organisation for the unique opportunities which it has afforded him for work in the field of cement-using roads and to his colleagues on the staff, whose enthusiasm, professional approach and willing assistance have been a constant source of inspiration.

Finally, it is too much to expect that work of this nature would be free of mistakes or bias. The author will be grateful to receive a note of any failings of this kind.

London SW1                                        D. RAYMOND SHARP

CHAPTER 1

# Concrete as a Material

## Cement

TYPES OF CEMENT

Cements used in highway engineering can conveniently be classified in two broad types: Portland cement, of which there are several varieties, and high alumina cement.

*Ordinary Portland cement (British Standard 12).* Ordinary Portland cement has a medium rate of setting and hardening and is used for most types of highway work.

*Rapid hardening Portland cement (British Standard 12).* Rapid hardening Portland cement, as its name implies, hardens more rapidly than ordinary Portland cement and is used when concrete of higher early strength is required. It is similar in chemical composition to ordinary Portland cement, but the proportion of the various component compounds may be slightly different and it is more finely ground. Rapid hardening Portland cement should not be regarded as a quick setting cement, since the term "hardening" refers to its rate of increase in strength and not to its setting time. The setting time specified in British Standard 12 for rapid hardening Portland cement is, in fact, the same as that specified for ordinary Portland cement. It should be noted that although at early ages the strength of concrete made with rapid hardening Portland cement is greater than that with ordinary Portland cement, after about six months both types of concrete have approximately the same strength.

*White and coloured Portland cements.* White and coloured Portland cements are used in highway engineering for delineating special carriageway areas such as marginal strips, parking areas, acceleration lanes,

## 2   Concrete in Highway Engineering

or for decorative structural purposes. White cement complies in all respects with British Standard 12 and is made from clays from which certain impurities are absent. Coloured cements are made by mixing a pigment with either grey Portland cement or white Portland cement. A wide range of the paler colours is obtained by using a white Portland cement base, while deeper shades such as buff, khaki and red can be made much more cheaply using ordinary grey cement as a base.

*Portland blast-furnace cement (British Standard 146).* Portland blast-furnace cement consists of a finely ground mixture of Portland cement clinker and selected granulated blast-furnace slag. The properties of blast-furnace cement are very similar to those of ordinary Portland cement, but it evolves less heat and is slightly more resistant to some forms of chemical attack and often gains strength more slowly at early ages.

*Low heat Portland cement (British Standard 1370).* Low heat Portland cement hardens and evolves heat more slowly than ordinary Portland cement. It is unlikely to be used in highway engineering except perhaps in very massive retaining walls where large volumes of concrete are required and where the heat of hydration cannot escape easily.

*Sulphate-resisting Portland cement (British Standard 4207).* Concrete made from sulphate-resisting Portland cement has a higher resistance than ordinary Portland cement to sulphates and is used in highway engineering where soils or ground water containing sulphates are in contact with concrete. It behaves similarly to ordinary Portland cement in regard to setting and hardening and the manufacturers should be consulted regarding the strength of sulphate concentration which it will resist in particular circumstances. For example, higher concentrations can be tolerated if the sulphate-bearing ground water is static rather than flowing when new solutions are constantly being presented to the face of the concrete. It should be noted, however, that an essential condition in the production of a concrete resistant to sulphates—and indeed to most aggressive environments—is that the concrete itself should be dense and of high quality. A lean, porous concrete will not give a durable concrete whatever type of cement is used.

*Hydrophobic cement.* Hydrophobic cement is ordinary Portland cement which has been treated so that it does not readily absorb water during storage. Its use in highway engineering is likely to be restricted to cases when its increased cost can be justified by this property of resisting deterioration under adverse storage conditions.

*Supersulphated cement.* Supersulphated cement is composed chiefly of granulated blast-furnace slag, gypsum and a small percentage of Portland cement. Concrete made from it is more resistant than ordinary Portland cements to sulphates in ground water or soils, but because of special requirements during storage and during the initial curing period it should be used strictly in accordance with the manufacturers' recommendations.

*Extra rapid hardening Portland cement.* Rapid hardening Portland cement containing an intimately mixed accelerator (calcium chloride) and known as extra rapid hardening Portland cement is used when high strengths are required at very early ages or when concreting in cold weather. The quicker evolution of heat resulting from accelerated hardening helps to prevent the concrete from being damaged by freezing temperatures during the critical early period after placing. It should be noted that extra rapid hardening Portland cement not only hardens more rapidly but also stiffens and sets more quickly than the cements already mentioned and care must, therefore, be taken to ensure that it is placed into its final position before this early stiffening action makes it difficult to achieve full compaction and a satisfactory surface finish.

*Masonry cement.* Masonry cement consists of Portland cement combined with a fine inert powder and sometimes a plasticising agent and it is used for rendering and block- and brick-laying. Its advantages are that it has a slower rate of hardening, a reduced strength and increased plasticity compared with ordinary Portland cement and so produces more satisfactory mortar because it is easier to use and has less tendency to crack.

*High alumina cement.* High alumina cement is quite different in composition from the Portland cements. It sets comparatively slowly, but hardens very rapidly with the production of a considerable amount of heat. It should not be used in rich mixes or in large masses of concrete and it is essential that the concrete be kept continuously wet for at least 24 hours after it has hardened. Care should be taken to ensure that structural high alumina cement concrete is at no time during its life allowed to rise in temperature above 85°F in damp conditions for an extended period. These conditions can bring about a change in composition and structure called conversion which can result in severe reductions in strength. High alumina cement concrete is more resistant than Portland cement concretes to the action of sulphates and to weak acids, but is liable to leaching and disintegration if continually exposed to alkaline solutions.

In highway engineering, the use of high alumina cement will be restricted to cases where it is essential for the concrete to achieve its full strength at an age of 24 hours. Because of the phenomenon of conversion, the use of high alumina cement concrete for pavement structural purposes requires careful consideration of the future environment of the concrete.[1, 2] In order to prevent the occurrence of a "flash" setting action Portland and high alumina cement concretes should not be allowed to come into contact with one another when they are both in the plastic state.

**Aggregates**

The maximum size of aggregate in general use for highway work is $1\frac{1}{2}$ in., although $\frac{3}{4}$ in. is now increasing in popularity. The term "coarse aggregate" is used to define material mainly retained on a $\frac{3}{16}$ in. B.S. sieve. "Fine aggregate" is the natural sand, crushed rock sand, crushed gravel sand or other material that mainly passes through a $\frac{3}{16}$ in. B.S. sieve. Gravels and crushed rocks such as granite, basalt and the harder limestones and sandstones are in common use as coarse aggregates for highway engineering. Natural sands are usually used for the fine aggregate in concrete, although crushed rock fines can be used for cement stabilised materials.

A list of rocks under their trade groups is given in Section 2 of B.S. 812, *Methods for Sampling and Testing of Mineral Aggregates, Sands and Fillers*. This British Standard describes tests for ascertaining whether aggregates are suitable for concrete, but the permissible limits for the test results are given in B.S. 882, *Specification for Aggregates from Natural Sources for Concrete*. Air-cooled blast-furnace slag used as coarse aggregate is dealt with in B.S. 1047, *Air-cooled Blastfurnace Slag Coarse Aggregate for Concrete*.

The two important characteristics in aggregates for concrete in highway work are durability and cleanliness. The standards required under these two headings differ, however, according to whether the concrete is exposed to the weather as in a bridge or road slab, or whether it is protected as in a lean concrete base. For concrete exposed to the weather or the action of traffic, aggregates should be hard and should not contain materials that are likely to decompose or change in volume when exposed to the weather. Examples of undesirable materials are coal, pyrites and lumps of clay. Coal may swell and disintegrate, pyrites may cause unsightly stains and lumps of clay may soften and form weak pockets. Again, when subjected to weather or traffic, all aggregates for concrete should be clean and free from organic impurities. Dirty aggregate containing vegetation, humus and other organic material weakens the concrete. The individual aggregate particles should be free from coatings of dust and clay as these reduce the bond between aggregate and matrix and reduce the strength of the concrete. Lower standards for aggregate can be accepted in cement stabilised materials for bases—for example, a higher percentage of dust or clay or soft particles can be accepted because high strengths in the base are not essential.

We are particularly fortunate in Great Britain in the wide range of good aggregates available for concrete. In some areas only one type of aggregate is available locally but in many areas the engineer will have a choice. In making his decision, the following factors should be taken into account.

1. Is the concrete to be used as a running surface? Some crushed rock fines are undesirable because they can produce concrete which often has a reduced ability to withstand wear and weather. Limestone fine aggregate produces slippery surfaces. Limestone coarse aggregate should only be used if information about the skidding resistance of concrete made with the material is available or on the basis of laboratory tests.

2. Can the concrete be readily finished with the plant to be employed to form a good riding surface? Some aggregates are better than others in this respect and a full scale trial using the plant should be undertaken before finally choosing the aggregates (see Chapter 4).
3. Is the concrete to be sawn in order to form joints? Some aggregates are more readily sawn than others (see Chapter 4).
4. Has concrete made with the aggregates under consideration a good local record for durability? Experience has shown that some aggregates produce concrete which tends to break down when exposed to freezing or wet conditions.

**Water**

The water specified for making concrete should be tested in accordance with B.S. 3148, *Tests for Water for Making Concrete*. The essential requirement is that the water should be reasonably free from impurities such as suspended solids, organic matter and dissolved salts which may adversely affect the properties of the concrete, especially the setting and hardening. The use of sea water does not normally affect the strength or durability of concrete, but it is not recommended in situations where efflorescence, should it occur, would mar the appearance of the work, as for example in bridge abutments.

**Concrete Mixes**

The engineer is concerned with the properties of concrete in both the plastic and the hardened states. When concrete is in the plastic condition it must be possible to mix, transport, place and compact it satisfactorily, and when it is hardened it must have the properties assumed for it in the design stage. The workability of the concrete is the property which determines the amount of work necessary to produce full compaction, that is the removal of the entrapped air, and is the most important property in the plastic state. In highway work the workability of concrete is usually measured by the compacting factor test,[3] although the slump test is sometimes used. Figure 1 shows the use of the compacting factor test to control the workability of concrete on a large concrete road project. The workability of concrete increases as:

1. The water content is increased.
2. The aggregate particles become rounder and smoother in texture.
3. The maximum size of the aggregate is increased.
4. The grading of the aggregates becomes coarser.

Concrete mixes can conveniently be designed by using the data given in Road Note No. 4, *Design of Concrete Mixes*.[4]

$1\frac{1}{2}$ in. maximum size aggregate is used for the more lightly reinforced structures, for concrete road slabs and for cement stabilised materials. $\frac{3}{4}$ in. maximum size coarse aggregate is used for the more heavily reinforced structural sections and for concrete road slabs, particularly for

Fig. 1. The compacting factor test being used on a site to determine the workability of the concrete.

the top layer in two-course construction. $\frac{3}{8}$ in. maximum size aggregate is used for small and heavily reinforced structural concrete sections and the design of these mixes is explained in Cement and Concrete Association Research Report No. 2.[5]

The grading of aggregates is generally continuous, that is particles of all sizes below the maximum are present. For certain purposes, however, gap gradings are used in which certain sizes of particles are entirely absent. For example, in mixes for concrete road slabs, the fraction between $\frac{3}{16}$ in. and $\frac{3}{8}$ in. is sometimes omitted, particularly if crushed rock is used, as experience has shown that this omission sometimes improves the ease of surface finishing. Again the omission of fractions between $\frac{3}{16}$ in. and $\frac{3}{4}$ in. is sometimes practised when special surface finishes are required. The ratio of fine aggregates, i.e. material passing through a $\frac{3}{16}$ in. sieve, to the total is important, particularly for concrete running surfaces (see Chapters 4 and 8). All-in aggregate, i.e. materials supplied containing all sizes of aggregate, should not be used for making concrete where uniformity of properties is desirable. These mixed materials vary in grading because they are often as-dug materials and also because they are liable to segregate during transport and handling. Reconstituted aggregates, i.e. aggregates with the full range of grading formed by adding together the required proportion of single size materials, are better in this respect, but they are still liable to segregate during transport and handling. Figure 2 shows an example of a $\frac{3}{4}-\frac{3}{16}$ in. aggregate which was supplied as "graded" but which has not been properly mixed. The use of two separated sizes, $\frac{3}{4}-\frac{3}{8}$ in. and $\frac{3}{8}$ in.–$\frac{3}{16}$ in., would be an improvement.

SEGREGATION OF CONCRETE

Segregation in a concrete mix occurs when some of the constituents tend to separate from the main mass, e.g. the larger particles tend to roll towards the edges if concrete is dropped into a heap from a height. Generally, segregation is indicative of poor aggregate grading, incorrect water content or faulty handling techniques and can usually be overcome. For example, segregation of coarse aggregate indicates a deficiency of fines. Gap graded mixes are particularly prone to segregate and their sand content is critical in this respect. Segregation of the coarse aggregate is also likely to occur in lean, dry mixes whilst separation of the cement and water paste can take place when too much water is used in the mix.

COMPACTION

Details of concrete strengths commonly quoted in specifications—for example, relationships between water/cement ratios and strength—apply only to fully compacted concrete. Concrete in this state is relatively easy to obtain in the laboratory, but it cannot be overstressed that the concrete mix must be designed to suit the compaction means available on the works site in order to produce the required degree of compaction or density. In structural concrete 5% of air voids reduces the strength by about 30% whilst the same percentage of air in a cement stabilised material can reduce the strength by half.

Compaction is the removal of the air entrapped by mixing and handling and is brought about by working the concrete. Much can be done in designing the mix to produce a concrete which is readily workable and yet which will possess the desired properties of strength and durability. In contrast, it is possible to design a very economical low

Fig. 2. A graded aggregate $\frac{3}{4}$–$\frac{3}{16}$ in. showing poor mixing. The use of two separate sizes—$\frac{3}{4}$–$\frac{3}{8}$ in. and $\frac{3}{8}$–$\frac{3}{16}$ in.— is preferable.

cement content dry mix which will produce the desired properties provided the correct amount of work is applied in compaction. If this amount of work is too great to be readily and universally applied to the concrete on site, not only will that concrete be of low strength but it will also be porous and lack durability when exposed to the weather. It should be borne in mind that it is often cheaper to increase the cement content of a mix to obtain easy compaction on the site rather than to devise a mix which requires very careful control of compaction on site coupled with specialised plant. This point particularly applies to mixes for structural purposes where the members are heavily reinforced.

## STRENGTH OF CONCRETE

Although concrete is stressed in flexure, in tension and in compression, the compressive strength is almost invariably used in practice in Britain as the measure of quality in the field. The compression test is simplest because the specimens are easy to make, are relatively small and the apparatus to test them is widely available. The testing of a cube in compression is often criticised on the grounds that the concrete is specially handled and does not necessarily bear any direct relation to the material in the structure and also because the test measures a property which is rarely called in question in practice. This widespread use of compression testing of cubes is not, however, as antediluvian as may appear at first sight. Most of the important properties of concrete are related to the compressive strength, and although not ideal it is probably the best test at present available. Most practising engineers have long experience of the test and the interpretation of the results. The greatest disadvantage is the lapse of time between making and testing the specimens and there is scope for development here and a new approach to concrete quality testing in general. Non-destructive methods of assessment of the strength of concrete exist,[6] e.g. ultrasonic testing and the rebound hammer,[7] but their relatively poor accuracy is a handicap to universal use. The cylinder splitting test is now being used in Britain and measures indirectly the tensile strength.

## DURABILITY OF CONCRETE

Most concrete used in highway work is subject to exposure to the weather to some extent but concrete road slabs have in addition to withstand abrasive wear caused by vehicles. The abrasive effect on good dense concrete caused by rubber-tyred vehicles is negligible and it has been found in practice that concrete designed to possess a minimum 28-day cube crushing strength of 4000 lb/in² is satisfactory in this respect. Some types of vehicles, e.g. tracked vehicles, do, however, have an increased effect and it is recommended that a concrete having a minimum 28-day crushing strength of 6000 lb/in² and made with a tough rock aggregate be used in such cases.[8] The criteria for the production of concrete with a high resistance to weathering can be summarised as good compaction without excessive surface laitance and low proportions of free water in the plastic concrete. Dirty and flaky aggregates and oversanded and overworked mixes must be avoided. The cement content of the mix as such is relatively unimportant as long as the concrete is fully compacted,[9] but a maximum aggregate cement ratio of 7 is probably advisable. The likelihood of frost damage can be reduced by using fully compacted concrete with a low water/cement ratio to produce a dense matrix with a pore structure in which a bursting tendency due to a build-up of harmful ice lenses will not occur. In recent years the removal of snow and ice from roads and paved areas has been increasingly effected by sprinkling calcium chloride or common salt onto the surface, a procedure which aggravates the damage produced by frost. Although concrete which will stand up to this procedure can be produced, an easier practical solution is to use an air entraining agent in the concrete as a safeguard against scaling.

ADMIXTURES

There is an increasing number and variety of admixtures on the market and it is not surprising that engineers are finding it a difficult task to assess their value to decide when their use is appropriate.[10] The most useful criterion to apply is that no admixture should be used unless it can clearly be shown that it will be advantageous. Admixtures should not be used to correct imperfections in mix, in handling or in placing the concrete. In many cases the effect of a particular admixture cannot be predicted accurately and it is, therefore, desirable to carry out preliminary tests under the actual site conditions to ensure that the improvement desired is in fact achieved and that no detrimental side effects result. The admixtures likely to be used in highway engineering are as follows.

1. *Accelerators.* Accelerators are used to speed the progress of concrete work or to enable work to proceed in cold weather. They do this by accelerating the hydration of the cement, thus producing more heat. The most common accelerator is calcium chloride and in commercial flake form this is used in a concentration not exceeding 2% by weight of the cement. Extra rapid hardening Portland cement can be obtained which achieves the same object and it has the advantage of convenience on site and uniformity of action.

2. *Air entraining agents.* Air entraining agents possess the property of incorporating non-connected minute bubbles of air into the concrete mix. The effect of this entrained air is quite different from that of entrapped air, a small amount of which is always present in plastic concrete even when compacted. The principal property of these agents is to increase the resistance of the concrete to frost. All concrete which is likely to be subjected to freezing and thawing cycles and to have neat salt applied to its surface should contain an air entraining agent. The other marked effects produced in concrete by entrained air are an increase in workability, a reduction in segregation, a reduction in strength and a reduction in bleeding at the surface. It follows that slight modifications have to be made in normal mix design procedure if these side effects are to be reduced to a harmless level.[11] The minimum amount of entrained air necessary to prevent frost damage to road concrete under conditions found in this country is probably about 4%. Specifications commonly call for $4\frac{1}{2} \pm 1\frac{1}{2}\%$ in order to allow a reasonable working tolerance.

3. *Water reducing or wetting agents.* The effects of water reducing or wetting agents is to increase the workability of a given mix without an increase in the water content. Their use is sometimes accompanied by a loss of strength and some agents entrain small quantities of air. If the workability is restored by reducing the amount of mixing water a net gain of strength can sometimes be obtained, and whether it is economical to use these agents in order to reduce the cement content requires investigation in each individual case.

4. *Retarders.* Retarders are mixed in with the concrete or used to

delay the hardening of a concrete surface. Integral retarders are used to delay the setting of concrete and they may also decrease the rate of development and ultimate strength. Some retarding agents also have the effect of reducing the amount of mixing water required. Retarders are used when placing conditions are difficult and when the mortar must remain plastic for a considerable time as in hot, dry weather or when large amounts of monolithically cast concrete are required as in some bridge piers. Surface retarders are used as a means of obtaining an exposed aggregate finish to concrete structures by painting a thin film onto the inner face of the formwork (see Chapter 8).

5. *Waterproofing agents.* Waterproofing agents fall into two groups, water repelling agents and pore filling materials. The most effective method of making impermeable concrete is to reduce the amount of water which it contains coupled with the use of a well-graded non-porous aggregate, the whole being fully compacted and adequately cured. Water repelling agents are added to the mix in order to reduce the capillary rise of moisture. It is doubtful whether they are effective under any hydrostatic pressure and their use usually results in loss of strength in the concrete. Pore fillers are materials which tend to reduce the pore size of concrete. They can be useful in producing a denser concrete if the aggregate used is deficient in fines. Neither type of agent will produce a concrete which resists the passage of water vapour.

## Hot and Cold Weather Concreting

Concreting work can be carried on in both abnormally hot and cold weather, although difficulties arise which are not present when temperatures are temperate. The measures necessary to cope with low or high temperatures are neither complicated nor very expensive, but in the case of freezing conditions a full economic study is necessary before deciding if the extra cost is worth while as compared with delaying the work until warmer weather returns.

### HOT WEATHER WORK

Problems arise when concreting in hot weather because of the higher temperature of the concrete and because of a greater tendency for the concrete to dry out as a result of evaporation of the mixing water. In Britain structural concrete members are unlikely to be adversely affected but relatively large exposed areas such as road slabs do sometimes suffer. Laboratory tests have shown that as the temperature of the freshly mixed concrete and the initial curing temperature increase there is a reduction in strength but it is unlikely to be significant in the conditions which are found in Britain. The practical difficulties of the reduced time for transporting and placing the concrete due to earlier stiffening in hot weather must, however, be taken into consideration. Measures which can be taken to reduce these difficulties include the following.

1. *Keeping down the temperature of the concrete by controlling the temperature of its ingredients.* The temperature of freshly mixed concrete can be estimated accurately from the temperature, specific heat and quantities of the ingredients.[12] Aggregate stock piles can be cooled by keeping them moist and so encouraging evaporation of water. The water has a significant effect on the temperature of concrete although it is used in small quantities, because its specific heat is about five times that of cement and of the aggregates. The water should be obtained from a cool source and should not, for example, be stored in tanks exposed directly to the sun. The use of crushed ice as part of the mixing water has proved highly effective in reducing the temperature of the concrete in extreme cases. Cement is often delivered to a site at a relatively high temperature and although the temperature of the cement produces significantly less change in the temperature of the fresh concrete than does that of the other ingredients, it does have some effect. Apart from this contribution to raising the temperature of the concrete, high cement temperature appears to have no detrimental effects on concrete.

2. *Reducing the time which elapses between mixing and placing concrete.* This period should always be kept to a minimum, but is even more important in hot weather. Loss of workability by evaporation of water occurs more rapidly in hot weather than at normal temperatures. Mixing should be timed and vehicles despatched so that delay will be avoided and work so organised that the concrete can be used promptly after delivery to the site.

*Concrete as a Material* 15

3. *Adding extra water to the concrete.* In hot weather a small amount of extra water compensates for the amount lost by evaporation during transport and placing.

4. *Arranging all placing and finishing procedures to keep the concrete as cool as possible.* This ensures that the concrete stiffens and hardens under reasonably uniform and moist conditions to minimise early drying out. The concrete should be finished as promptly as possible. Cooling formwork, reinforcement and the subgrade with water just prior to placing concrete is helpful but care should be taken to avoid using standing water. The final finishing of road slabs should be carried out as soon as possible after the compaction process. In extreme cases the surface of the slabs can be sprinkled with water from a fine jet in order to aid the finishing process but care should be taken to use the absolute minimum quantity of water. This practice is not to be encouraged because it can result in the presence of a layer of weak mortar at the surface, but it is nevertheless probably preferable to the alternative of an open surface of aggregate particles incompletely surrounded with mortar.

5. *Paying special attention to protection and curing.* These operations are more critical in hot than in cool weather, since hot weather leads to rapid drying of concrete. In extreme cases water should be applied to the surface of concrete even whilst the forms are in place. Curing of exposed surfaces should be started as soon as possible, preferably immediately after the final finishing. Wet curing or complete covering with waterproof paper or plastic film should be continued for at least 24 hours and preferably longer. Beyond this period white pigmented curing compounds can be used.

6. *Using retarders to delay the setting time.* This is a particularly useful procedure when casting large masses of concrete where it is desirable to avoid hardened joints. Retarders should only be adopted where the concrete is subject to a good degree of control.

PLASTIC CRACKING IN CONCRETE

Cracking that occurs in the surface of fresh concrete soon after it has been placed and before it has hardened fully is called plastic cracking. The cracks appear mostly on horizontal surfaces and may be eliminated if appropriate precautions are taken. The principal cause of plastic cracking is rapid drying of the concrete at the surface, and even when the same materials, proportions, and methods of mixing, handling, finishing and curing are used cracks may develop intermittently due to a change in weather conditions which varies the rate of evaporation from the surface. If the rate of evaporation exceeds the rate at which water rises to the surface, then plastic cracking is likely to occur on the concrete immediately after compaction. These cracks are usually quite shallow and soon diminish in width and are not often detrimental except to the appearance. They can be avoided by improved curing methods and by delaying the final surface finishing so that they are closed up.

COLD WEATHER CONCRETING

In most parts of Britain freezing temperatures seldom continue for more than a few days, so it is usually more convenient and economical to stop paving concreting when the thermometer drops than to take the necessary precautions to permit concreting to continue. Concrete can, however, be laid satisfactorily in freezing conditions provided the concrete itself is prevented from freezing before it has hardened sufficiently and experience gained in recent severe winters has shown that no great practical difficulties arise. Concrete which has reached a crushing strength of 750 $lb/in^2$ is unlikely to be damaged by frost.

The precautions to enable concreting to continue in cold weather are all designed to prevent the water in concrete from freezing and to obtain 750 $lb/in^2$ crushing strength as soon as possible. If the water contained in the fresh concrete freezes, the mass is expanded so that normal strength can never be reached and disintegration usually results. The precautions to be taken depend on the severity of the weather and for this purpose cold weather can be divided into three categories:

1. When the temperature is below 40°F but does not fall below freezing point.
2. When frosts occur only at night and are not very severe.
3. When night frosts are very severe and frost continues all day or nearly all day.

Concrete as a Material 17

Table 1 summarises briefly the precautions to be taken in each case. As temperatures are so important when concreting in cold weather, a thermometer must be used to record them. A maximum and minimum thermometer to register air temperatures in the vicinity of the work is desirable and, in addition, a maximum and minimum thermometer

TABLE 1. PRECAUTIONARY MEASURES IN COLD WEATHER CONCRETING

| 1<br>Always take these precautions at low temperatures | 2<br>Add these when there is slight frost at night | 3<br>Add these also if there is severe frost day and night |
|---|---|---|
| Extend the curing time and keep the formwork in position longer | Make sure that the aggregates are not frozen | Heat the water and aggregate |
| and | and | |
| Insulate the concrete after placing to prevent loss of heat | Make sure that surfaces in contact with the concrete are not frozen | Mix the concrete near the job |
| or | | |
| Accelerate hardening by<br>(a) using an extra rapid hardening, or rapid hardening cement, or<br>(b) using an accelerator | | Enclose the placed concrete and provide continuous heating |
| and | | |
| Insulate the concrete | | |

should be placed immediately adjacent to the concrete surface where it is most likely to be affected by frost. For example, in the case of road slabs a maximum and minimum thermometer should be placed in contact with the concrete surface underneath the protecting insulation. Detailed procedures for concreting in cold weather both for structural work[13, 14] and for concrete road slabs[15] have been published. The

18  *Concrete in Highway Engineering*

use of extra rapid hardening cements and accelerators, both useful aids to cold weather concreting, has already been mentioned earlier in this chapter. Further information is given in Chapter 9.

Finally, it should be remembered that although it is quite feasible to concrete in cold weather and prevent damage to the concrete, the rate of gain of strength of concrete is retarded by low temperatures. Near freezing point the strength gain is very low and at temperatures below freezing there is almost no increase in strength. Concrete cured in air at just above freezing point, for example, will have a 28-day strength which is only about half of that of the same concrete kept at 80°F.

**Curing Concrete**

The chemical action of setting and hardening of concrete can only proceed in the presence of water. Normally there is an adequate quantity available in the mix for full hydration, but it is necessary to ensure that this water is either retained or replenished to enable the chemical action to continue until the concrete is fully hardened. The purpose of curing is to maintain the concrete in a continuously moist condition over a period, either by preventing evaporation and absorption of water by forms or by the subgrade, or by repeatedly wetting the surface to replace lost water. Wet curing appears to give the best environment for concrete, and the nearer the curing conditions approach this state the more impermeable, durable and strong will be the resulting product. The length of time moist curing is necessary is dependent on the type of cement used and the temperature at which curing takes place. Concrete made with ordinary Portland cement, blast-furnace or sulphate resisting Portland cement should be cured for at least 7 days under average conditions, but this period can be reduced to 3 days if rapid hardening Portland cement is used. Supersulphated cement concrete requires special curing conditions. When using any cement in cold winter weather, the normal curing period should be doubled and the length of time during which the temperature is below freezing point should be added. Curing should begin as soon as the concrete is sufficiently hard for its surface not to be marked by the covering or by the spray.

The methods of curing in practical use are dictated by the necessity

Concrete as a Material 19

for economy and by the type of construction. The methods commonly used include the following:

1. Spraying with water.
2. Covering with wet hessian or matting which is kept continuously wet.
3. Covering with wet sand which is kept continuously wet.
4. Covering with waterproof paper or plastic sheeting which will prevent evaporation of water from the concrete.
5. Flooding with water by building small dams, for example of clay, around the area to be cured.
6. Spraying with sealing or "membrane" compounds. White pigmented compounds are more efficient in sunny weather as they reduce the heat intake. These compounds are best applied to the surface just as the sheen of water is disappearing.

If concrete cannot be covered immediately after finishing it must be protected from the sun and drying winds by means of tents of fabric supported just clear of the surface and closed around the edges to prevent the evaporation of water by draughts as far as possible. In the case of road slabs these tents can conveniently be supported on frameworks mounted on wheels.

### References

1. NEVILLE, A. M. A study of deterioration of structural concrete made with high-alumina cement, *Proceedings of the Institution of Civil Engineers*, vol. 25, paper no. 6625, July 1963, pp. 287–324.
2. INSTITUTION OF STRUCTURAL ENGINEERS. *Report of a committee on the use of high-alumina cement in structural engineering*, London, August 1964, pp. 17.
3. BRITISH STANDARDS INSTITUTION. B.S. 1881, *Methods of testing concrete*, London.
4. ROAD RESEARCH LABORATORY. *Design of concrete mixes*, Road Note No. 4, London, H.M.S.O., 1953, pp. 16.
5. CEMENT AND CONCRETE ASSOCIATION. *The workability of concrete mixes with $\frac{3}{8}$ in. aggregates*, Research Report No. 2, London, June 1955, pp. 7.
6. JONES, R. *Non-destructive testing of concrete*, London, Cambridge University Press, 1962, pp. 104.
7. KOLEK, J. An appreciation of the Schmidt rebound hammer, *Magazine of Concrete Research*, vol. 10, No. 28, March 1958, pp. 27–36.
8. COLLINS, A. R. and WATERS, D. B. The resistance of road surfacings to tank traffic, *Proceedings of the Institution of Municipal Engineers*, vol. 72, No. 6, January 1946, pp. 221–240.

9. KEENE, P. W. *Some tests on the durability of concrete mixes of similar compressive strength*, Cement and Concrete Association, TRA 330, January 1960, pp. 18.
10. SHACKLOCK, B. W. The use of admixtures in concrete, *Concrete and Constructional Engineering*, September 1962, p. 8.
11. SHACKLOCK, B. W. Air entrained concrete—properties, mix design and quality control, *The Surveyor & Municipal & County Engineer*, vol. 119, No. 3560, 27 August 1960, pp. 969–971.
12. CEMENT AND CONCRETE ASSOCIATION. *Hot weather concreting*, Advisory Note No. 10, CZ 10, London, May 1958, pp. 7.
13. CEMENT AND CONCRETE ASSOCIATION. *Concreting in cold weather*, Man on the Job Leaflet No. 19, London, July 1955, pp. 11.
14. PINK, A. Winter concreting, *Civil Engineering & Public Works Review*, vol. 60, No. 710, September 1965, pp. 1331–1335.
15. CEMENT AND CONCRETE ASSOCIATION. *Guide to concrete road construction*, 2nd edition, London, November 1958, pp. 74.

CHAPTER 2

# Pavement Design

IN this context of design the term "pavement" will be used to denote the whole road structure excluding earthworks. The word, although widely used, is an unhappy choice, if only because of the risk of confusion when it is used to denote a footpath, but unfortunately no suitable alternative exists.

**Types of Pavement**

Pavements are divided into two types, rigid and flexible. These are bad descriptions in so far as they postulate a method of behaviour which is inaccurate. For want of better alternatives, however, they have been used in this text.

Concrete is the only important form of rigid pavement. It has considerable powers of bridging over small weaknesses and depressions in the subgrade soil. Flexible pavements are unable to resist small tensile stresses and any change in the shape of the surface of the subgrade is followed by a corresponding change in the road structure. The traditional form of flexible pavement is a bituminous surfacing laid in one layer or more on a base and sub-base of compacted mechanically stable materials. Cement stabilised base and sub-base materials such as lean concrete and soil-cement are not fully rigid but for convenience of description are commonly classed as flexible materials. They do, however, possess considerable load spreading powers. The basic structural elements of concrete and flexible pavements are as set out in Figs. 3 and 4.

## Principles of Design

A pavement is built up of several layers, each having a special function. In a flexible system the load is largely distributed by the base, the main function of the surfacing being to provide a wearing surface and to protect the base. The sub-base, although it has to carry a smaller intensity of load, is nevertheless important but is usually composed of material of lower quality.

In a concrete road the slab usually provides the wearing surface as well as spreading or distributing the load. A sub-base is almost always

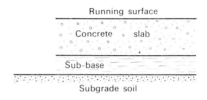

FIG. 3. Concrete road terminology.

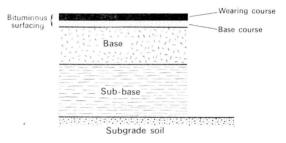

FIG. 4. Flexible pavement terminology.

used under a concrete slab as it serves as a useful working platform from which to construct the slab. Concrete roads are sometimes surfaced with a bituminous carpet, particularly in urban areas.

The object of pavement design is to determine the type, thickness and treatment of materials that will most economically provide an adequate wearing surface and structure to carry a given frequency and weight of vehicles on given foundation conditions. The strength of the subgrade soils commonly found in Britain varies greatly even over short distances,

Pavement Design 23

as do the weight and intensity of traffic using our roads. It is debatable whether it is sound economics to design a pavement to be strong enough for the worst possible combination of these circumstances. A considerable saving in cost arises if, say, 1 in., representing perhaps as much as 15% of the total, can be cut off the thickness of a pavement. A portion of the road may fail due to inadequate thickness but this may still be an economical approach if repairs are easy and cheap. A second point is that a failure in a road is acceptable because, unlike the case of a structure, it does not give rise to immediate danger to life. Although the engineer may be able to show that it is sound economics to accept a failure of a percentage of a new road, it is doubtful if the public would accept this approach. The danger of failure can be reduced by adequate expenditure on preliminary site investigation and the intelligent use of realistic design methods.

**Traffic Loading**

In Britain, Road Note 29 prepared by the Road Research Laboratory has for some years been the standard work of reference for pavement design.

Traffic is usually considered in terms of the number of commercial vehicles per day as private vehicles cause very little distress except for very light roads.

A commercial vehicle is defined as a goods or public service vehicle of unladen weight exceeding 30 cwt. Design methods usually take account of the normal number of notifiable heavy loads as defined by the Road Traffic Acts. Where the number of such loads is likely to be unusually heavy—for example, on a road adjacent to a gravel pit or docks—discretion should be used in choosing a design corresponding to a higher traffic category. For the purpose of design, an estimate is made of the probable number of commercial vehicles expected to be using the road during its life. In making this estimate account should be taken of the existing traffic data, including the effect of replacing existing roads or new construction in the vicinity. The traffic information which will normally be available from census data is an estimate of the initial traffic at the time of construction expressed in commercial vehicles per day and an estimated growth rate. Where no estimate of growth can

24  Concrete in Highway Engineering

be made, an average value of 4% per year can be taken. Curves are available in Road Note 29 giving the cumulative number of commercial vehicles carried by each slow lane for various design lives and for different traffic growth rates. The average number of axles for commercial vehicles at the present time is assessed and using this factor, the cumulative number of commercial vehicles can be converted to a cumulative number of axles. From a knowledge of axle load spread by weight on typical roads, the cumulative number of commercial axles can be converted to standard axles in tons of damaging power using axle load equivalent factors.

**Design of Flexible Pavements**

There are a great number of methods available for designing flexible pavements, but in Britain the California Bearing Ratio (C.B.R.) method is now almost universal. Briefly, the test, which is an arbitrary comparison, consists of forcing a cylindrical plunger 3 in.$^2$ in end cross-section into a sample of soil at the rate of $\frac{1}{2}$ in./min and measuring the load required to cause a penetration of 0·1 in. or 0·2 in. This load is expressed as a percentage of the load obtained when using a standard material. Full details of the test are readily available.[1] The application of the test to design is based on results obtained over a number of years by the Californian State Highway Department on materials comprising the subgrade, sub-base and base of roads that had both failed and survived. These tests lead to the conclusion that a material with a certain C.B.R. required a certain minimum thickness of construction above it. The intensity and weight of traffic using these roads were investigated and this led to the production of various design curves for different wheel loads. These wheel loads were later interpreted by the Road Research Laboratory in terms of numbers of commercial vehicles using a road daily and the results put in graphical form for practical use.[2] It should be noted that the C.B.R. test is not practicable if the material to be investigated contains an appreciable amount of material coarser than $\frac{3}{4}$ in. If this is so the suitability of this material for use has to be assessed visually.

When designing new pavements the main difficulty in using the C.B.R. method is to decide under what conditions of moisture content and dry density to test the subgrade in order to allow for changes subsequent to construction. In the U.S.A. the test specimens are usually soaked before making the test but this is probably too unrealistic and could lead to serious overdesign. A more rational approach is to carry out the C.B.R. test on soils preferably in the undisturbed condition at the highest moisture content which they will attain in practice. This may well be the equilibrium moisture content, which is the moisture content of the subgrade soil when it has been covered by the overlaying pavement and moisture conditions have become static, but it may be the anticipated moisture content immediately after construction.

If the pavement is constructed in the summer or if the area is in heavy cut, it is likely that the moisture content will increase after construction until the equilibrium value is reached. Alternatively, if construction is carried out in winter it is likely that the subgrade soil will decrease in moisture content to the equilibrium value. The estimation of these various moisture contents is not always easy, but much can be done by experience and experiment. For example, the equilibrium moisture content can often be ascertained by tests carried out on soils under an adjoining paved area, which need not, of course, be a road. If no direct method is possible, the equilibrium content can be assumed as being equal to the moisture content of the soil between 3 and 4 ft down.

Although the thickness of any individual layer of material comprising a pavement designed by the C.B.R. method is dependent on the C.B.R. of that particular material, the present day tendency is to use surfacing and base materials which possess high and uniform C.B.R. values and to use them at standard thickness dependent upon the intensity of traffic.[1] The sub-base material is specified as having a minimum C.B.R. and the thickness varies according to the C.B.R. of the subgrade. Minimum thicknesses of 4 in. to 6 in. are normally required for bases and sub-bases because of the practical difficulty of laying thinner layers.

The question of frost susceptible subgrades needs special consideration. It has been ascertained that the average maximum depth of frost penetration in Britain in bad winters is about 21 in. and a total construction depth of at least 18 in. is often called for if roads laid in frost susceptible subgrade materials are not to be adversely affected. Measurements during the winter of 1962/3 have shown that frost can penetrate up to 26 in. Soundly constructed pavements with impervious surfaces did not, however, suffer damage even though their total construction depth was no more than 18 in.

Due to the inadequacies of the C.B.R. test it is probable that flexible pavement design in Britain will in future be based on a simple classification system of subgrade soil and drainage condition for which standard proved pavement systems are chosen depending mainly on traffic intensity.

The prepared formation of a new road should be protected from the action of weather prior to the construction of the pavement. The finished surface of any layer making up the pavement should be similarly protected if it is formed of a material which is likely to deteriorate if it dries out or wets up. A simple form of protection is a surface dressing.[3]

CEMENT STABILISED MATERIALS

The C.B.R. method of test was designed for unbound materials relying for stability on internal frictional forces. It is to be expected that the load bearing capacity of concrete stabilised materials will be correspondingly greater and it follows that the C.B.R. method of design will give extravagant thicknesses. Many authorities have recognised this and have reduced the base and sub-base thickness of cement stabilised materials given by C.B.R. methods by 15% to 25%. On the other hand, many authorities prefer to look upon the increased strength and stability obtained from cement stabilised materials as being a bonus to guard against future eventualities, including unexpected traffic increase, and they therefore design to full C.B.R. thickness. No more precise advice can be given in the present state of knowledge but when the results of full-scale experiments now being conducted by the Road Research Laboratory are available, more exact data should be forthcoming for use in this country.

The most recent AASHO Road Test[4] in America clearly showed the superiority of stabilised materials over materials laid as a mechanically stable base. It is already clear from these experiments that cement stabilised materials behave excellently in every respect as long as they are not overstressed. Once they do reach this condition, their failure is likely to be rapid unless the constituent aggregate is mechanically stable.

## Pavement Design    27

It follows, therefore, that it is undesirable to reduce the thickness of cement stabilised materials to the absolute minimum. No expansion or contraction joints are normally provided in cement stabilised pavements and construction joints are simple butt joints.

**Design of Concrete Pavements**

In addition to the normal design factors of thickness of slab, type and thickness of sub-base, strengthening of edges and corners of slabs and use of reinforcement, it is convenient to include the type and spacing of joints under the heading of pavement design of concrete.

The thickness of slab required for a given site will depend on:
(a) the properties of the soil,
(b) the intensity and weight of traffic,
(c) the amount of steel reinforcement present, if any,
(d) the thickness of the base.

These four factors will be considered in greater detail.

(a) *Properties of the subgrade soil.* It can be shown by theoretical calculations using the formulae due to Westergaard that the bearing capacity of the soil has little effect on the stresses in the concrete when slabs are uniformly supported, i.e. when the soil has a uniform bearing capacity. In practice, however, uniformity of support is rarely obtained due to the variable nature of soils and to changes in moisture content during and after construction. In addition non-elastic deformation of the soil occurs due to traffic. From the practical evidence available and also from the results of experimental roads, it appears that the same thickness of slab is suitable for most types of subgrade. Present day design procedure in Britain recognises only three different types of soil, which are designated as "normal", "very stable" and "very susceptible to non-uniform movement". In practice probably over 80% of concrete roads are built on "normal" subgrades.

(b) *Intensity of traffic.* Experience has shown that the intensity of traffic has by far the greatest effect on the performance of concrete roads but that light vehicles such as private cars have practically no effect structurally. In design tables, therefore, the number of commercial

Fig. 5. A crack in an unreinforced concrete slab showing excessive width and incipient deterioration due to edge spalling.

vehicles using the roads is the only factor taken into account. As it is not practicable or economic to strengthen a concrete slab later in life to take account of increasing traffic loads and intensity, it is particularly important to anticipate the type of traffic which the road will carry throughout its life, as described earlier in this chapter under **"Traffic Loading"**.

(c) *Steel reinforcement.* Experimental work has shown that the use of reinforcement does not permit the slab thickness to be reduced, its main effect being on the slab length which can safely be laid without harmful cracking. The AASHO test[4] is a valuable reference in this

Pavement Design    29

context. The spacing between free joints in reinforced slabs is calculated from the concept that the amount of steel available at any cross-section must be sufficient to withstand the tensile stresses compounded from the weight of the slab and the friction between the base and underside of the slab when a crack has occurred and when the concrete is contracting. The spacing of joints in reinforced pavements is dependent, therefore, upon the thickness of the slab, the value of the friction coefficient and the cross-sectional area of reinforcement in the slab.

Experience has shown that in unreinforced slabs, joints must be placed at a distance not exceeding between 15 and 20 ft if cracks are to be avoided. The use of reinforcement therefore allows joint spacings to be increased. It should be stressed that it is very desirable to avoid cracks in unreinforced slabs. If these occur they are likely to increase progressively in width and to spall with subsequent fairly rapid deterioration under heavy traffic. Figure 5 shows a crack in an unreinforced slab; it will be noted that it is comparatively wide and that edge spalling has begun to occur. Cracks in reinforced slabs are not serious because the reinforcement prevents the cracks from opening, thus reducing the chance of water penetration and of spalling. Experience shows that cracks in reinforced roads suffer very gradual deterioration and often remain unchanged in width for many years.

It follows from the concept of the action of reinforcement given above that the maximum advantage is obtained if the greater proportion of steel is placed in the direction of the greatest dimension of the slab. A rectangular mesh is therefore appropriate and the transverse bars are normally limited to those required to space the longitudinal bars effectively during handling.

Reinforcement used in road slabs is almost always specified to be in accordance with B.S. 1221 and because of its convenience in handling, fabric made from hard drawn steel wire by electrically welding the transverse and longitudinal bars together is most popular. The weights of reinforcement given in concrete road design tables are based on the use of this type of reinforcement.

Reinforcement should always be delivered to sites in flat mats and not in rolls as use of the latter causes difficulty in accurately placing the steel.

The use of two layers of reinforcement makes construction more difficult and tends to cause difficulties in obtaining full compaction in the concrete, particularly under the lower layer. As there is no convincing evidence that better performance is obtained with two layers compared with one layer with the same total weight, the use of a single layer of fabric placed about 2 in. from the top of the slab is now almost universal. The only exception is when heavy weights of reinforcement are being used structurally—for example, over peat subsoils when positive and negative bending moments have to be designed for.

(d) *The provision of a sub-base.* Experimental work has indicated that sub-bases fulfil very little purpose in strengthening a concrete road structure which obtains its load carrying capacity mainly from the structural rigidity of the slab. In general, experiments have indicated that 1 in. of concrete is equivalent to about 6 in. of sub-base and unless suitable sub-base material can be obtained very cheaply it will usually be economical to increase the thickness of the slab rather than that of the sub-base in order to increase the traffic carrying capacity of the road. In practice, a sub-base is used under a concrete slab mainly for construction purposes, that is to protect the subgrade soil and to facilitate the movement of construction traffic. There are, however, one or two exceptions. Some clays and silts, for example, are liable to exhibit the phenomenon called "mud pumping", that is they readily assume the consistency of mud when mixed with water and under this condition are sometimes pumped out through joints or cracks in concrete slabs as a result of the action of traffic. On clays and silts, therefore, a sub-base is essential in order to prevent mud pumping unless traffic is light. Again, it may be necessary to provide a sub-base to insulate frost susceptible subgrade soils from frost.

Access to roads under construction is rarely easy unless traffic is allowed to run on the area of the carriageway itself. Although attempts have been made to carry out construction working from the sides, these have proved only partially successful, and it is probably more economic and certainly more realistic to allow construction traffic on the sub-base. Newer design methods recognise this and permit reduction in slab thickness when the sub-base is strengthened for construction traffic.[5]

On highly plastic subgrade soils a sub-base thickness of at least 6 in. is necessary in order to prevent construction traffic from damaging the formation. Where construction traffic is likely to be heavy—for example, on motorway work—this thickness may have to be increased, although it depends to some extent on the choice of material. Thus, a lean concrete or soil-cement sub-base will more readily carry construction traffic in all weathers without deformation than a mechanically stable granular material like hoggin.

CORNER AND EDGES OF SLABS

It is readily apparent that for any given load the edges and corners of a concrete slab are stressed more highly than the middle portion. If a balanced design is to be produced, it follows that the edges and corners should be strengthened. This can be done by thickening the outer edges of slabs, constructing concrete sleeper beams under joints and edges of roads or by increasing the amount of reinforcement at the edges.

None of these methods has proved entirely satisfactory or advantageous in practice and the best method of coping with the problem is probably to prevent traffic running over the edges of slabs and by providing load transfer devices at joints so that slabs give support to one another. Traffic can be prevented from running on the edges of slabs by placing kerbs so that the kerb is about 12 in. from the edge of the slab. This expedient cannot be adopted, however, when a hard shoulder is placed adjacent to the slab but the problem is probably less acute in this case because the subgrade soil conditions are likely to be good at the slab edge. Concentrations of water at the edge of the slab with subsequent risk of softening of the subgrade soil at this danger spot are prevented as the discharge of surface water is carried well clear of the slab. Widening the sub-base to at least 12 in. beyond the edge of the concrete slabs is of considerable assistance in providing full support to the slab edges and also in protecting the subgrade soil from moisture changes. Extra reinforcement is sometimes placed at edges of slabs, notably in Switzerland, and may be effective in reducing any tendency for cracks to commence at these edges.

Corner reinforcement or hairpin reinforcement is sometimes used in thin slabs which are subjected to heavy loads—for example, 6 in. slabs as overslabbing to existing concrete roads. Inside edges of slabs constitute longitudinal and transverse joints and except on lightly trafficked roads they should be provided with load traffic devices such as tie bars and dowel bars to provide mutual support between slabs.

JOINTS

Transverse joints are provided in concrete roads to allow for expansion, contraction or warping of the slabs caused by changes in temperature and moisture content of the concrete and to allow for a break in construction at the end of the day's work. Longitudinal joints are required to allow for warping in wide slabs and to allow the road to be laid in lanes of convenient width.

Joints must fulfil a number of functions. They must permit slabs to move without restraint but they must not unduly weaken the road structurally. They must be effectively sealed so as to exclude water and fine matter such as grit, and yet must not detract from the appearance or riding quality of the road.

*Types of transverse joints.* The transverse joints should be arranged in a continuous line across the width of the road, all joints in this line being of the same type. Staggered joints can give rise to sympathetic cracking in the slabs opposite the joints. Joints should be at right angles to edges of slabs. Experiments in this country in the use of transverse joints at angles to the slab edges have not indicated any advantage either in riding quality or in performance, although they are used in some states of America.

Transverse joints include expansion joints, contraction joints, warping joints and construction joints.

*Expansion joints.* These consist essentially of gaps between slabs to allow for expansion (see Fig. 6). The gaps are usually filled with compressible material in order to prevent the ingress of foreign matter. Expansion joints have always traditionally been used in this country but for some years in countries such as America, Belgium, France, Holland, Switzerland and Denmark they have been omitted without detrimental effect. The risk of compression failure, buckling or "blow-ups" is slight provided any construction joints are carefully formed so that they are vertical and provided no weak parts are present in the slabs as a result of variation of thickness or inadequate compaction of the concrete. Indeed, in one warm summer in Belgium no buckling failures occurred in roads laid without expansion joints, although a number were experienced in roads with expansion joints, which had become filled with grit over the years, causing eccentricity. Roads without expansion joints have been laid in Britain on a trial basis for some years without difficulty and Road Note 29 now permits roads to be constructed in the summer without provision for expansion.

*Contraction joints.* These are breaks in the continuity of the concrete slab permitting it to contract with a drop in temperature. Contraction joints also permit some angular movement, so allowing for warping which occurs under the effect of temperature gradients between top and

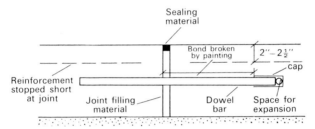

FIG. 6. Expansion joint with dowel bars in reinforced slab.

bottom of slabs (Fig. 7). Contraction joints extend the full depth of the slab or consist of a groove about one-quarter of the slab depth formed in the top of the slab either when the concrete is plastic or after it has hardened. Contraction causes a crack to spread from the bottom of the groove downwards, thus relieving tensile stresses.

*Warping joints.* These are similar to contraction joints except that they do not allow for opening of the joint. Tie bars or reinforcement are carried across the joints which allow a small amount of angular movement, so preventing the development of high stresses due to restrained warping. Warping joints have not been much used in Britain, although they have found favour in Holland and Switzerland.

*Construction joints.* These are formed when work has unexpectedly to be interrupted at a place where no joint would otherwise be required or at the end of the day's work. Normally, however, the day's work should be so planned that it ends at a joint required for structural purposes, e.g. at a contraction joint. Construction joints are fully tied, often with reinforcement carried through and supplemented by tie bars, the object being to prevent movement or opening.

*Load transfer devices.* In order to reduce the stresses at transverse joints and in particular at expansion joints, it is advisable to strengthen them by incorporating load transfer devices which share the loads between adjacent slabs. Experience shows that by far the most satisfactory method for expansion and contraction joints is the use of dowel

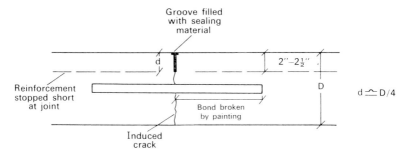

FIG. 7. Contraction joint with dowel bars in reinforced slab.

bars, which are a series of steel bars bonded into one slab and sliding in the other. A typical section of dowel bars in a contraction joint is shown in Fig. 7. Dowel bars are not recommended in slabs less than 6 in. thick because cracks tend to appear over them, due to inadequate cover of concrete. For their proper performance it is essential that dowel bars shall be able to slide freely and that they are parallel with one another and with the surface and longitudinal axis of the slab.

*Longitudinal joints.* Longitudinal joints are required in concrete roads more than about 15 ft wide, not only for construction purposes in semi-manual work, but also to allow for transverse warping and for uneven settlement of the subgrade. They can extend the full depth of the slab or consist of a groove about one-fourth of the slab depth formed

## Pavement Design 35

in the top of the slab either when the concrete is plastic or when it has hardened (Figs. 8 and 9). Load transfer devices are desirable in longitudinal joints to prevent relative settlement and to reduce the stresses from vehicles. Tie bars are commonly used for this purpose and

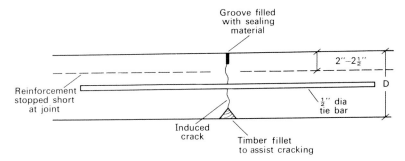

FIG. 8. Longitudinal joint with tie bars in reinforced slab cast full width.

these consist of mild steel bars bonded half in one slab and half in the other. Tie bars differ from dowel bars in that they prevent relative horizontal movement of adjoining slabs. Simple butt joints are generally

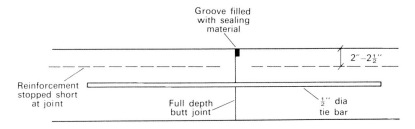

FIG. 9. Longitudinal joint with tie bars in reinforced slab cast in strips.

used for full depth longitudinal joints although tongue and groove joints are sometimes preferred. Thorough compaction of the concrete in the tongue and groove is, however, essential and because of the difficulty of guaranteeing this, the value of their use is problematical.

## 36 Concrete in Highway Engineering

### DESIGN OF PAVEMENT THICKNESS, JOINT SPACING AND REINFORCEMENT WEIGHT

Designs for concrete pavements including slab and sub-base thickness, the weight of reinforcement if used and joint spacing for different traffic and subgrade conditions are given in Road Note 29.[1]

The choice of a design with expansion joints only or one with both expansion and contraction joints is normally based on economics and consideration of the class of work involved. On the smaller job, the use of two kinds of transverse joints gives more complications and many engineers and contractors will prefer the simplicity of a design which includes all expansion joints and all contraction joints only.

The design of road pavements differs from that of structures in the sense that structures are designed to be safe for an indefinite period. Road pavements, on the contrary, are designed with a definite life in view although highway engineers are usually not clear as to the exact figure which they should aim at or are indeed aiming at. Experience indicates that well constructed concrete roads to present designs should last 40 years before they need major remedial measures. Even then the life of the slabs is not necessarily at an end because they can be surfaced with a bituminous carpet. If a shorter life is required, e.g. for a quarry access road, the designs can safely be reduced but each case must be treated on its merits.

### DESIGN OF CONCRETE BASES UNDER BITUMINOUS SURFACING

There will be occasions when it will be desirable or necessary to surface a concrete slab with a bituminous carpet, e.g. in city streets or to match existing work, and the design of the slab must then be varied. If expansion, contraction or warping joints are incorporated in a slab which is surfaced, the movement which occurs at these joints will almost certainly be transmitted through the carpet and appear at the surface as a crack. This can be avoided if the slab is constructed with continuous reinforcement without any joints. Construction joints will, of course, be necessary but these should be fully tied by continuing the reinforcement through or by using tie bars or preferably both.

Slabs constructed in this way will quite clearly crack due to shrinkage and thermal movements but the magnitude of the cracks will be small and their frequency great because of the presence of reinforcement. The chance of the crack being reproduced, therefore, in the surfacing is slight, particularly if thick carpets are employed. There are many examples of roads constructed in this way in Britain, including experimental roads,[6] and experience has been uniformly satisfactory. Compressive stresses are built up in the concrete during warm weather unless the carpeting material is very thick but in a well constructed slab consisting of well compacted concrete these stresses can readily be sustained.

Experiments have shown that carpet thicknesses greater than 2 in. have an insulating effect but below this thickness the amount of solar radiated heat absorbed by the concrete with its black surfacing is likely to be greater than the unsurfaced light coloured concrete. Surfaces thinner than 2 in. have, however, been successfully employed in practice, but they are not recommended except for lightly trafficked roads where good riding quality is of lesser importance.

## References

1. ROAD RESEARCH LABORATORY. *A guide to the structural design of flexible and rigid pavements for road-work*, Road Note No. 29, London, H.M.S.O., 1965.
2. ROAD RESEARCH LABORATORY. *Soil mechanics for road engineers*, London, H.M.S.O., 1952, pp. 541.
3. ROAD RESEARCH LABORATORY. *Protection of subgrades and granular bases by surface dressing*, Road Note No. 17, London, H.M.S.O., 1962, pp. 4.
4. SHARP, D. R. and SHACKLOCK, B. W. *A British assessment of the AASHO road test with special reference to concrete pavements*, Technical Report TRA 369, London, Cement and Concrete Association, February 1963, pp. 17.
5. SHARP, D. R. *Design of concrete and cement stabilized roads*, Fifth World Meeting of the International Road Federation, London, September 1966, Cement and Concrete Association, pp. 8.
6. BROOK, K. M. and PULLAR-STRECKER, P. J. E. H. *Behaviour of a continuously reinforced concrete road at Dundee*, Technical Report TRA 356, London, Cement and Concrete Association, March 1962, pp. 15.

CHAPTER 3

# Subgrade Soils, Sub-bases and Drainage

**Moisture Control and Drainage**

It has long been appreciated that the stability of a road pavement is enhanced if the subgrade soil remains in a dry, uniform condition. This condition is impossible to achieve in a climate as wet as that of Great Britain and the best the highway engineer can do is to reduce the

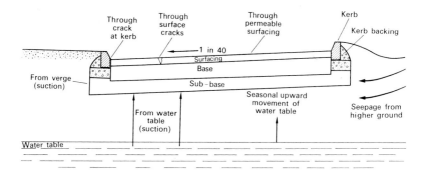

Fig. 10 Possible movements of water into a road structure.

chances of changes in moisture content in the subgrade and to design his pavement thickness for the worst case—that is, for the highest moisture content likely to occur in practice.

Changes in moisture content in the subgrade can occur for a number of reasons. These include seepage from higher ground, rise or fall in the level of the water-table, percolation of water through from the road surface, and transfer of moisture to or from the verges or from the lower soil layers (Fig. 10). All these can be minimised by various expedients, the most important of which is drainage. The main object of drainage is to stabilise the soil moisture conditions so that the strength of the soil will be as high as possible and as uniform as possible. As discussed in Chapter 2, the strength of the subgrade soil directly affects the thickness of flexible pavements. Concrete pavements are not so dependent upon the bearing capacity of the subgrade, but they are sensitive to variations from place to place locally.

**Control of Surface and Subsoil Water**

A completely impermeable road surface is difficult to maintain in practice, and similarly it is almost impossible to prevent some subsoil water from reaching the pavement structure. There are two fundamental approaches for dealing with the problem. The first presupposes a porous sub-base consisting of compacted stable material such as a crushed stone, which is placed on a carefully cambered or shaped formation and connected to a drainage trench. This porous sub-base disperses any water passing through the layers of the pavement placed above it and at the same time harmlessly traps any water reaching the pavement from the sides. As it is made from stable open graded material, it adds some strength to flexible pavements and will effectively prevent the phenomenon of mud pumping under concrete slabs. It also protects the subgrade soil to some extent from damage by construction traffic although softening can occur as rain passes easily through the sub-base.

In practice, porous sub-bases are difficult to achieve in this country because of the scarcity of suitable material to construct them. In addition, experience has shown that they are not fully effective for long periods. For this reason, sub-bases made of impervious material have usually been chosen. As-dug or reconstructed gravel and sand mixtures are the most common choice but crushed stone, crushed concrete, slag and cement stabilised soil have also been used.

The stabilisation of the top 3 to 6 in. of the subgrade with cement is probably the most effective method of providing a stable sub-base. Construction traffic can safely use this layer, although its thickness might have to be increased to 8 in. or 9 in. if traffic is heavy. Once it is

Fig. 11. Drainpipe laid in hard shoulder to dispose of surface water trapped on top of cement stabilised base prior to surfacing. Note entry holes for water in kerb.

laid the rest of the pavement can be constructed in good working conditions. The cost of stabilisation may well be recouped from the saving which results from the fact that the contractor can move about the site and construct the pavement in wet weather. One carriageway only need be used for construction traffic and the sub-base on the second can therefore be constructed to lower standards. Care must be taken to control surface water during the construction period. Dangerous quantities of water trapped in the road structure during construction have led to premature failure. Figure 11 shows a method used on one site to dispose of surface water trapped on a base by a concrete margin before the bituminous surfacing was applied.

**Frost Susceptible Soils**

The construction of roads on frost susceptible subgrade soils poses special problems. These soils "heave" or expand when ice forms in the upper layers. Figure 12 shows a flexible road constructed on chalk during a period of prolonged frost. The formation of ice lenses has caused uplift of the whole road structure as shown by the depression at the manhole cover which, being supported on the shaft, has not moved. Damage occurs when the ice melts because traffic loads then have to be carried by the soil in a wet disturbed condition. In Great Britain the greatest depth of frost penetration is up to 26 in. or more, and so a construction depth equal to this must be provided if all possibilities of frost heave on frost susceptible soils are to be removed. Except in the case of very important or heavily trafficked roads, however, some risk of frost heave can be taken, in which case the total construction depth can be reduced (see Chapter 2). Concrete roads, to judge by their performance in the bad winters of 1947 and 1963, appear to be less affected by frost heave of subgrade soils and settle back undamaged, particularly if all joints are dowelled. All materials used in the construction of the pavement itself must, of course, in themselves be completely frost resistant.

**Layout for Subsoil Drainage**

The digging of trenches under pavements is to be avoided whenever possible as it disturbs the subgrade soil, reduces its bearing capacity locally and introduces non-uniformity. For this reason, pipe drains are usually laid in the central reservation or beyond the edges of the carriageway. Before designing a subsoil drainage system it is necessary to establish the conditions with which it will have to cope and this is

FIG. 12. Frost heave in a flexible pavement constructed on a chalk subgrade. The manhole cover has remained at the original level.

best done by a survey of the soil and the moisture conditions in the subgrade. The following information is relevant:

1. The soil type and thickness of the various strata.
2. The position of the water table.
3. The position of seepage zones which are likely to affect the pavement.

If seepage is present within 2 or 3 ft of the surface of the subgrade then an intercepting drain can be laid on the high side of the carriageway.

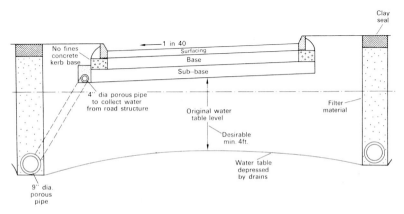

FIG. 13. Lowering the water table level by installing drains.

If the seepage zone is wide, it is generally possible to keep the seepage water about 4 ft below formation level by constructing intercepting drains. A high water table can also be lowered by the installation of a drainage system (Fig. 13). The actual spacing and depths of drains will depend on the soil conditions and width of the formation. For example, in the case of dual carriageways, drains may be necessary under the central reservation as well as under the edges of the formation.

In a drainage trench a pipe provides a convenient channel for the removal of the drained water whilst the trench itself provides the greater part of the drainage action. Porous pipes are commonly used and the drainage material filling the trench should be so graded that it resists the ingress of fine material which would soon reduce its efficiency.

Porous concrete pipes are widely used because they do not become blocked as a result of clogging of the pores by fine material. It is good practice when using filter materials to seal the top of the drainage trench as the surface run-off usually contains some silt in suspension which tends to cause silting up of the back fill. A suitable seal is a layer of compacted puddle clay.

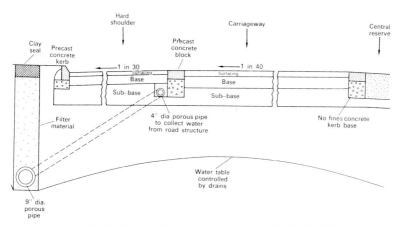

FIG. 14. Drainage of dual carriageway with hard shoulders (flexible construction).

## Layout for Surface Water Drainage

Roads provided either with crossfall or camber have both proved satisfactory in disposing of surface water to one or both edges when the carriageway width is limited to about 30 ft. Serious problems arise on three-lane and wider carriageways laid to a single crossfall because the resulting high concentration of surface water along the pavement edge is undesirable both from the point of view of traffic using the road and because there is more risk of water penetrating into and underneath the road structure. The adoption of fully paved hard shoulders on motorways and other important roads has accentuated the problem. Experience in other countries suggests that these undesirable factors can be eliminated or reduced by using a depressed central reservation and cambered carriageways. With this design, drainage from the central reservation onto the carriageway surface is eliminated and run-off from the carriageway lanes is split and flows to both edges. The necessary cambers can conveniently be formed using straight lines instead of curves in major road construction.

Recent experience of wide carriageways has also shown that the practice of permitting surface water to trickle onto the soil at the carriageway edges is not entirely satisfactory. Positive collection devices such as kerbs, gulleys, channels and ditches are necessary. Where a hard shoulder is provided a channel and kerb placed on the outside are effective and are essential on embankments to prevent local erosion (Fig. 14).

Similar difficulties arise in controlling any water which might be present in or immediately under the pavement. For this reason, subbases which are continuous over the whole width of the pavement from one outside edge of the shoulder to the other, including the central reservation, are finding favour.

Water sometimes finds its way into a road structure although the amount involved will be small if the surface and the component layers are nominally impervious. Means of escape should be provided for this water by the provision of a drain at the pavement edges or by using no-fines concrete in the bottom portion of the concrete haunch where one is provided.

CHAPTER 4

# Concrete Roads

**History**

The concrete road is one of the oldest forms of construction still in large scale current use. Although basically the same as the original concept, the modern variation, due to refinements over the years, is a very different product compared with its predecessor of even two decades ago. When concrete was first used for roads it was thought that once the hard weather-resistant covering of concrete was laid, everybody's worries would be at an end. The years have shown the fallacy of this principle and it is now realised that the preparation of the sub-grade soil and the design and construction of the sub-base under the concrete are of primary importance if the potential long life and good performance of the superimposed slabs are to be realised.

Concrete roads are not as popular in Britain as their past performance record would justify, and this is due to several reasons. The most important is that the very long life, which is one of the attributes of this form of pavement, can react against the laying of new concrete roads. This arises because both engineers and the lay public tend to judge the performance of concrete roads, and in particular the riding quality, by the condition of existing roads, forgetting that these roads have deteriorated over the years. It is, moreover, forgotten that they did not even when new possess the high standard of riding quality which is judged to be essential in the context of the present day. It is demonstrably unfair to use the yardstick of a 30–40-year-old road to judge construction carried out now with the full knowledge of research results and of experience.

## The Economics of Concrete Roads

Concrete roads are often rejected because it is claimed that they are too expensive. It is true that in some areas concrete is more expensive than the cheapest alternative, but this is only part of the story. The life of the concrete running surface is considerably greater than that of any bituminous surfacing. An investigation carried out by the Road Research Laboratory[1] has shown that the concrete roads surveyed, which were of pre-war and early post-war construction, had an average life of $27\frac{1}{2}$ years. Modern roads will undoubtedly improve on this figure. The amount of money spent in maintenance of these roads was rather less than $\frac{1}{2}$d. per sq. yd per year.

When concrete slabs reach the end of their useful life as a running surface they can then form the base—which will be virtually everlasting —for a bituminous surfacing. It is sometimes claimed that the cost of applying this bituminous surfacing will be excessive because of the amount of deformation which will have taken place in the concrete slab, necessitating considerable repair work followed by the application of a bituminous surfacing at least 3 in. thick. This was true of roads constructed to older designs, but present-day roads constructed without expansion joints should perform much more advantageously under bituminous surfacings because they will not require extensive treatment at joints in order to prevent the reproduction of cracks.

The present-day national financial policy for roads encourages a low initial cost, irrespective of whether the form of construction chosen will result in excessive annual maintenance. Added to this, the methods of keeping records in use by local authorities do not encourage the analysis of annual costs of different types of road as opposed to their initial costs. It is certain that many local authorities would, if they kept and analysed their records accurately, find that concrete roads are considerably cheaper than any alternative form of construction on an annual cost basis. A further advantage in favour of concrete arises because the small amount of maintenance work necessary results in turn in reduced dislocation to the highway using public and a saving in money to the community.

Blake and Brook[2] have analysed the various operations for major reinforced concrete roads. They concluded that the proportions of cost due to labour, plant and materials are 10%, 15% and 75% respectively. Expressed as a percentage of the total construction costs, the cost of aggregates, steel and cement are respectively 26%, 20% and 20%.

**Construction of Concrete Roads**

It is convenient to consider construction methods for concrete roads under three headings:

(a) Small-scale road work.
(b) Minor roads involving appreciable lengths where speed of completion is a factor.
(c) Major roads including motorways where speed of construction and high standards of work are of paramount importance.

A distinction is drawn between the methods of construction in use for the categories above, although originally all roads were constructed using the same hand methods. Full mechanisation first came into use in this country in the early 1930's, and the extent of its application and the machinery used and its refinement have been a continuing process ever since.

**Small-scale Projects—Semi-manual Methods**

SUB-BASES

For the reasons already given in Chapter 2, a sub-base under concrete roads is only necessary from the design aspect on very poor soils to protect the subgrade from frost. A sub-base is usually desirable in practice, however, to provide a hard, smooth layer over the formation upon which to work. It is prepared to an accurate surface level (often $\pm \frac{1}{2}$ in.) to ensure a uniform thickness of the slab. In addition, the sub-base will protect the subgrade soil from damage during construction of the concrete slab. Hoggin, well graded gravels or sands, lean concrete, cement stabilised soil or artificial materials such as clinker or slag can be used as sub-base materials. Hardcore may also be used, but it should be crushed so that it is graded from 2 in. down; it should not contain rubbish such as plaster. Ashes are unsuitable as they are unstable when wet. The top surface of the sub-base should be finished

Fig. 15. Checking the accuracy of the sub-base with a scratch template, using the forms as a datum.

and rolled to a smooth level surface, fine material being added if necessary to close the voids. Figure 15 shows the accuracy of finish of a sub-base on a small project being checked with a scratch template. It may sometimes be necessary to allow traffic on this sub-base before the concrete slab is laid. In this case the sub-base should be surface dressed to prevent surface damage and the admission of water.[3]

DESIGN OF THE CONCRETE MIX

On most small-scale projects it is desirable that the method of mix design for the concrete should be simple, as work of this type may be supervised by men without knowledge of the more sophisticated aspects of concrete technology. The use of arbitrary mix proportions to enable the required minimum strength at 28 days to be obtained is advocated although on larger sites the mix should be designed to suit the particular circumstances.[4] Suggested mixes which may need adjustments on individual sites are given in Table 2.

Experience has shown that concrete having a minimum cube crushing strength of 3000 $lb/in^2$ is all that is necessary for minor projects and the mixes given should give this figure at 28 days with ordinary Portland cement. On the small job, the strength may be neither specified nor measured, but the concrete should have this minimum strength so that satisfactory slab performance will be obtained. Experience has shown that if concrete of this strength is used, other essential requirements

TABLE 2. ARBITRARY CONCRETE MIXES FOR SMALL-SCALE ROAD PROJECTS

| Maximum size of coarse aggregate (in.)<br>Slump (in.)<br>Compacting factor | | $1\frac{1}{2}$<br>$\frac{1}{2}-1\frac{1}{2}$<br>0·85–0·90 | | | $\frac{3}{4}$<br>$\frac{1}{2}-1\frac{1}{2}$<br>0·85–0·90 | | |
|---|---|---|---|---|---|---|---|
| Fine aggregate zone (B.S. 882) | | 1 | 2 | 3 | 1 | 2 | 3 |
| Dry weights of aggregate (lb) per 112 lb cement for air entrained concrete | sand<br>coarse | 325<br>400 | 300<br>425 | 275<br>450 | 275<br>350 | 250<br>375 | 225<br>400 |
| Dry weights of aggregate (lb) per 112 lb cement for plain concrete | sand<br>coarse | 300<br>375 | 275<br>400 | 250<br>425 | 250<br>325 | 225<br>350 | 225<br>350 |

such as resistance to wear and to weather will also be obtained. The average strength of concrete needs, of course, to be higher than the minimum figure given above to allow for variations in materials and mix proportions. Table 2 makes an appropriate allowance for these factors in accordance with the standard of quality control usually found in this class of work.

Well made, dense concrete is highly resistant to the effects of frost, but it is nevertheless recommended that an air entraining agent should be used in the concrete mix, particularly if neat salt is likely to be applied to the surface for de-icing purposes. The total average air content required to suit British conditions is about $4\frac{1}{2}\%$ with a margin of $\pm 1\frac{1}{2}\%$ to allow for site variations. The quantity of air entraining agent used should be as advised by the manufacturers.

Weighbatching is recommended even for the smallest job because it gives a mix of precise proportions which can easily be checked. Modern small weighbatching plant is now freely available, either built into concrete mixers or as a separate piece of equipment. The water/cement ratio by weight should not exceed 0·55 for plain concrete to ensure satisfactory durability, but up to this limit the amount of water to be added is determined by the criterion of correct workability. Higher water/cement ratios can be used with air entrained concrete,[5] but a maximum of 0·60 is probably desirable.

The workability of the concrete must be high enough to ensure that a fully compacted material is obtained and to give a smooth surface finish on the slab surface. If the concrete is too dry and compaction inadequate, the slab will be honeycombed and will not be durable. If, however, the concrete is too wet, the slab will have a low strength and the surface will be liable to become smooth and to wear rapidly under traffic. The workability, however, must not be so high as to cause the concrete to flow down the crossfall or camber during compaction. Experience has shown that the workability of concrete for semi-manual work should vary between 0·85 and 0·90 compacting factor ($\frac{1}{2}$–$1\frac{1}{2}$ in. slump) depending upon weather conditions. Warm, drying weather requires the use of more workable concrete.

CONCRETE CONTROL TESTS

The routine tests normally made on concrete for road work are for workability, strength and air content if air entraining agents are used. The compacting plant itself applies a check on workability. The water content should be kept to a minimum, consistent with obtaining a fully compacted slab and a closed surface free from excess moisture. Ideally, the surface should come to a good finish just after the concrete reaches full compaction. If the size of the job warrants the making of cubes, single specimens should be made with concrete from six different batches at random every day during the first week. Once it is reasonably certain that the quality is right it should be sufficient to make four cubes a week.[6] Cubes should be tested at 28 days, but on small sites where the work is likely to be completed in less time the results of the strength tests will not be known in time to allow adjustments to be made to the mix. The cubes are, therefore, only used as a check on quality and it must be emphasised that the real control must be applied to the batching and compaction of the concrete. The air content of air entrained concrete should be checked frequently during the initial stages of the work and later twice-daily spot checks only will be required. A pressure type meter which measures the amount of water required to replace the air in the concrete is usually used for site work.

BATCHING, MIXING AND TRANSPORTING OF CONCRETE

Concrete can be batched and mixed either in central fixed plants or in machines which are moved along periodically to keep pace with the work, or ready mixed concrete can be used. Concrete is transported to

Fig. 16. Concrete batching and mixing plant showing cement silo, weighbatching equipment and timber separating walls between different aggregate sizes.

the placing site either in lorries, dumpers, truck mounted mixers or agitators. Figure 16 shows a typical mixer and general arrangements for the smaller concreting site.

READY MIXED CONCRETE

Concrete can now be purchased ready mixed in most parts of the country. It is made either in central batching and mixing plants and

Fig. 17. A small concrete job in progress using ready mixed concrete. The base course concrete and reinforcement have been placed in the foreground and the top course is now being spread and compacted. Note the boxing out around the manhole cover.

transported to the site in mobile agitators or tippers or in central batching plants used in conjunction with truck mounted mixers. The use of ready mixed concrete is particularly valuable on the small or restricted site and when the supply needed is intermittent. Figure 17 shows concrete being delivered to such a site.

As ready mixed concrete is made in purpose designed permanent plants by specialists, the quality is usually consistently high. It is, however, necessary to ensure that the supplier is provided with all the information he needs about the concrete required so that he can design an appropriate mix.[7]

## CONSTRUCTING THE CONCRETE SLAB

The continuous method of construction in which work proceeds continuously throughout the day is advocated, any joint separating materials and load transfer devices being incorporated in the concrete as the work proceeds. Alternate bay construction, that is leaving the intermediate bay to be filled in after the concrete of the first and third bays has hardened sufficiently, is not advisable because of the difficulty of controlling surface rain water.

Except where there is a hard smooth surface to the sub-base, waterproof paper or plastic sheeting should be used under the concrete slab to prevent the escape of cement paste into the sub-base and to reduce the friction between the slab and sub-base.

Steel forms are now invariably used and they should be fixed rigidly to the sub-base with an adequate number of steel pins with the locking devices driven fully home. The forms should be well supported, preferably on a continuous bed of cement mortar or concrete. Forms should be aligned over long lengths in order to ensure accuracy, and should be laid at least one day ahead of concreting. Any individual form which departs from the true profile by more than $\frac{1}{10}$ in. should be rejected as the shape of the form decides the accuracy of the running surface. Figure 15 shows a method of using the forms to check the sub-base.

On the smaller project the concrete will usually be placed by hand. The practice of dumping the concrete from the transporting vehicle on to the formation and levelling it in readiness for compaction should be avoided, as it causes high spots in the finished surface. The concrete should therefore be tipped on to a banker board either to the side or on the formation and the material shovelled into position by hand, thus giving concrete of uniform initial compaction. To allow for subsidence which occurs during compaction the concrete must be spread to a surcharge. The amount of this extra thickness will depend on the concrete mix and generally will be about one-fifth of the depth of the concrete slab when compacted. A light screed working from timber battens clipped to the top of the forms simplifies the striking off to uniform surcharge height.

The concrete slab is sometimes tamped off kerbs which are placed before the concrete is spread, but this is not practicable if artificial

Fig. 18. Concrete supplied ready mixed to a small job and compacted by a hand-guided vibrating beam. The scraping straight edge removes any remaining irregularities.

falls to gullies are necessary. Where a high standard of riding quality is required tamping should be off forms.

The concrete is compacted by a hand-guided vibrating beam. Pneumatic, electric or petrol driven vibrating units mounted on wooden or steel beams are used for the purpose. It is important that the beam is allowed to sink down to form level under its own weight and then lifted up and replaced on a fresh area of uncompacted concrete. The beam should not be allowed to move along under the propelling action of the vibrators. Although this procedure may produce a satisfactory surface, the concrete will not be densely compacted and the slab will be weak. In reinforced concrete slabs, concrete is spread in two layers to enable the reinforcement to be placed accurately. It is not always essential to compact the two layers of concrete separately but two-layer compaction is usually necessary in slabs 9 in. or more thick.

Finishing should be carried out continuously over as long a length of concrete as possible. Mechanically driven vibrators can be used for finishing the surface as well as for compaction. The finishing pass should be made soon after the compacting operation and the beam should be tilted slightly so that the leading edge is about $\frac{3}{8}$ in. higher than the trailing edge to facilitate the striking off of any small quantities of excess concrete.

*Concrete Roads* 55

It is often difficult to remove all irregularities in a concrete road with a hand-guided vibrating beam, but the use of a scraping straight edge will ensure a surface well within the usual specified tolerance of $\frac{1}{8}$ in. in 10 ft. The scraping straight edge consists essentially of an 8–10 ft long rigid metal or timber strip fitted with a long flexible handle.[8] Figure 18 shows a concrete slab being compacted and finished on a housing site.

The time allowed to elapse between mixing the concrete and final finishing should not exceed $1\frac{1}{2}$ hours. The most satisfactory final surface is obtained by brushing the concrete with a medium-hard broom after the final pass of the vibrator or scraping straight edge. The brush should be drawn across the road surface in one continuous sweep, producing a uniform texture which is skid resistant and which in-

FIG. 19. Forming a lightly brushed texture on a housing estate concrete road.

Fig. 20. Constructing an expansion joint on a housing estate road. Note the temporary supports for the dowel bars and reinforcement laid on the lower layer of concrete.

creases the light reflection at night (Fig. 19). Deeply indented ridged surfaces are not necessary except on very steep slopes. The construction of a concrete road using the methods described above is illustrated pictorially in the Cement and Concrete Association's publication *Housing Estate Roads*.[9]

## JOINTS

Expansion joints, contraction joints, longitudinal joints and construction joints are those most likely to be used in minor concrete roads. Expansion joints are provided to allow the concrete to expand beyond its original length. Figure 20 shows an expansion joint with dowel bars

Concrete Roads 57

under construction. Contraction joints prevent or control cracking caused by tensile stresses in the slabs when expansion joints are widely spaced. A longitudinal joint will be necessary where the slab is more than 15 ft wide. Construction joints should only be made when unavoidable interruptions occur in the progress of the work. Generally expansion joints or contraction joints only will be used on small jobs where labour is likely to be less experienced and the standards of supervision not very high.

An important introduction into this country has been the technique of sawing joints in concrete roads after the concrete is hardened. Water-cooled circular blades of various types are used. The engine driven mechanism and blade are mounted on wheels in a convenient portable frame which can be propelled along the road surface. The saw can be used for forming a sealing groove at a longitudinal joint, for forming a contraction joint, and also for sawing the groove to contain the joint sealing compound over the pre-formed jointing material in an expansion

FIG. 21. A semi-automatic concrete saw cutting a contraction joint groove in hard concrete.

## 58 Concrete in Highway Engineering

joint. The great advantage of sawing joints is that the riding quality is unimpaired by the presence of the joint and the evenness of the surface is not dependent on manual skill. Figure 21 shows a contraction joint sealing groove being sawn.

### CURING

On smaller sites curing is carried out by spraying with a proprietary membrane curing compound, by spreading plastic sheeting or other waterproof covering on the surface or by watering. Curing should continue for seven days in warm weather but in cold weather this period should be extended. The process should commence as soon as possible after the concrete has been laid and no slabs should be left unprotected after the original water sheen has evaporated, this being particularly important in warm or windy weather. Covering is also desirable to protect the surface from being pitted by heavy rain which might fall before the concrete has hardened.

### LAYOUT OF BAYS AT
### ROAD JUNCTIONS, GULLIES AND MANHOLES

At road junctions, islands, ends of culs-de-sac, etc., joints must be placed not only to allow for expansion and contraction in the concrete but also so that they can be constructed conveniently.[10]

When designing the slab shapes the following points should be considered:

1. Suitable falls to gullies should be provided and sudden changes in level avoided by the use of appropriate transition bays in approach roads.
2. There should be no acute angles between bays of concrete. Joints should be located so that joint intersections do not form angles of less than 90°.
4. Junctions should be constructed wherever possible without the use of special forms and in bays that can be properly compacted.
4. Gullies and manholes should be cast in small independent slabs to facilitate their removal and replacement as shown in Fig. 22.

*Concrete Roads* 59

OPENING THE ROAD TO TRAFFIC

In warm weather normal traffic can be allowed on the concrete road after 7 days if rapid hardening Portland cement is used and after 10 days if ordinary Portland cement is used. In cold but not frosty weather these periods should be increased by 50% and in a long spell of very cold weather with frosts the periods must be still further increased. Light rubber tyred traffic can be allowed to use the road after shorter periods.

## Medium-sized Projects

The methods of construction described under **"Small-scale Projects"** produce of necessity limited outputs and some degree of mechanisation must be carried out if greater daily production is required. The first step, assuming that increased concrete production capacity is available, is to speed up the compaction and finishing processes. For this purpose

Fig. 22. Boxing out of gullies and manhole covers in separate slabs. Note the well sell sealed joints.

Fig. 23. A simplified form of concrete finishing machine suitable for small-scale work.

machines such as that shown in Fig. 23 are available. These are simple forms of finishing machines, sometimes self-propelled and sometimes winched along the forms or kerbs, embodying a simple form of strike off device, vibrating beam and sometimes a finishing screed. A further step in the process of mechanisation is to use a machine for spreading, and a simple hydraulically operated blade spreader is shown in Fig. 24. This machine can deal with concrete only after it has been tipped on the formation and therefore has some limitations as it does not eliminate troubles due to precompaction of concrete at the bottom of the tipped heaps, although skilful use of the blade can help. Final finishing of the surface follows the procedure described for semi-manual methods.

The machines described here merit greater use than they have received in Great Britain as they can produce high quality work quickly and at a competitive price.

**Large-scale Projects**
BASES

Materials used for sub-base construction in fully mechanised methods include stable granular materials such as hoggin, crushed concrete,

Concrete Roads 61

soil-cement and lean concrete. Materials which lose a substantial part of their strength when they become wet will obviously be less suitable to carry construction traffic under wet conditions and there has therefore been a continuing trend to use higher quality sub-base materials, sometimes stabilised with cement, which are insensitive to changes in climatic conditions.

A second approach is to forbid the contractor to use the formation and sub-base for his construction traffic and to confine him to the central reservation and the hard shoulder areas. Experience of this procedure has, however, shown that it is unrealistic to confine the contractor in this way both from the time and from the economic standpoints. It has, however, been shown that it is sufficient to provide an all-weather sub-base on one carriageway only, enabling the concrete slab to be readily constructed first on this carriageway, the completed pavement then forming a sound sub-base from which to construct the second concrete carriageway. On large-scale motorway projects it is becoming increasingly common to construct the sub-base using fully mechanised methods and, in particular, to require a high standard of regularity and finish on the surface. Planing and trimming machines as shown in

FIG. 24. A mechanical spreader for use on small-scale concrete paving work.

Fig. 25. A wire-guided grader for trimming earthworks, sub-bases or bases to fine limits.

Fig. 25 are now in use. Some types are guided automatically from stretched piano wires.

DESIGN OF THE CONCRETE MIX

On larger projects the design of the concrete mix is usually carried out by the contractor in order to satisfy certain basic specification requirements. These in Britain are commonly:

1. The grading of aggregates shall be within the limits of B.S. 882, *Concrete Aggregates from Natural Sources*, and the grading zone of the fine aggregate once determined and approved shall not be changed without permission.
2. The maximum size of aggregate shall be either $1\frac{1}{2}$ in. or $\frac{3}{4}$ in.
3. A stated aggregate/cement ratio shall not be exceeded (7:1 by weight is common).
4. Ordinary Portland cement is normally used.
5. An air entraining agent is normally used with a total air content within the range of $4\frac{1}{2}\% \pm 1\frac{1}{2}\%$.
6. The minimum 28-day crushing strength of 6 in. cubes shall be 4000 lb/in$^2$.
7. The concrete shall be of suitable workability for full compaction to be obtained with the equipment used and without undue flow.

Once the optimum value of workability has been determined for the mix and plant being used, it should be maintained within a tolerance of $\pm 0.03$ compacting factor for uniform results. In most cases the same mix is used throughout the depth of the slab, but in others—for example, when cheap limestone is available which is not suitable for a wearing surface—a second type is used for the base course, that is beneath the level of the reinforcement when used. Richer, more workable mixes with sand contents up to 40% or more are sometimes used for the top course as experience has shown that they can be more speedily and readily finished to give the desired surface regularity.

CONCRETE CONTROL TESTS

The following tests of the concrete are those usually carried out on large-scale work:

1. Compressive strength of concrete (after 7 and 28 days).

Fig. 26. A large-scale concrete batching and mixing plant with an 8 cu yd mixer and aggregate stockpiles for a motorway contract.

Fig. 27. A travelling paver mixer feeding a spreader by moving belts on a large-scale motorway contract. Note (1) lorries bringing dry batched concrete for mixer, (2) reinforcement laid ready for placing, (3) plastic film laid on base, (4) tents covering spreading and finishing machines for protection from the weather.

2. Air content of concrete.
3. Grading of aggregates.
4. Workability of concrete.
5. Measurement of surface irregularity.
6. Cores may be cut when the quality of the concrete in the pavement as opposed to that in cubes is in question or when slab thicknesses have to be determined or when the degree of compaction being obtained is in doubt.

BATCHING, MIXING AND TRANSPORTING

Concrete can be batched and mixed in central fixed plants or dry batched in a central plant and mixed in a mobile mixer usually called a paver mixer which accompanies the train of construction machinery. Each method has its own enthusiasts; paver mixers are relatively uncommon in Europe but much used in America. Batch mixers are the most common type, although continuous mixers have also been used. Ready mixed concrete has not as yet proved very popular, mainly because of the difficulty of achieving a large enough output for the purpose. Outputs of 1500 cu yd and more of concrete per day can be required when fully mechanised construction methods are used to produce the full outputs of which they are capable. Concrete is transported almost universally from central plants to the construction site in tipping lorries. If mobile paver mixers are used, these lorries are usually provided with a number of compartments each holding enough dry batched concrete for one mix and these compartments are emptied in turn into the receiving hopper. Figure 26 shows a central batching and mixing plant whilst Fig. 27 depicts a site where mobile paver mixers are in use.

CONSTRUCTING THE CONCRETE ROAD

The basic operations of constructing slabs consists of laying the forms, spreading the concrete, compacting and finishing the concrete and curing it. The forms used include a rail of suitable weight to withstand the applied weight of the plant and these forms are set up one or two days in advance of the work. They are permanently bedded on concrete or cement mortar in order to facilitate their setting and to

FIG. 28. Laying a heavy duty form used for fully mechanised construction on a bed of plastic fine concrete.

Fig. 29. A box spreader receiving concrete from a side tipping lorry.

prevent movement in use (see Fig. 28). It is obviously desirable that the forms shall be of stout construction, undamaged and laid true to line and level as the accuracy of the finished surface depends upon them. Specifications usually require that forms shall be individually correct on the top surface within a tolerance of $\pm \frac{1}{10}$ in., although the overriding requirement is that the surface finish of the concrete shall be constructed accurately—normally not less than $\pm \frac{1}{8}$ in. tolerance under a 10 ft straight edge.

The type of spreader in most common use consists of a hopper mounted longitudinally in relation to the slab, which is filled either directly from a side tipping lorry or indirectly through a moving belt (Fig. 29). The best types are provided with a gate at the bottom of the hopper, which is advantageous as it enables the flow of concrete to be controlled. The aim of good spreading is to achieve a uniform thickness of uncompacted concrete.

FIG. 30. A movable belt concrete spreading machine placing concrete on a 38 ft wide carriageway slab. Note the reinforcement mats stocked ready for use in the foreground.

FIG. 31. An articulated concrete finisher fitted with an angled finishing beam.

FIG. 32. A concrete train constructing reinforced concrete slabs, the concrete being side fed from the adjoining carriageway.

The slab is compacted usually by vibrating beams, although poker vibrators mounted in line are sometimes used. The reinforcement is usually placed 2–3 in. from the running surface and duplicate spreaders and finishers are common for the bottom and top course, although smaller jobs have been constructed using one of each. Figure 30 shows a high output spreader working on a motorway site.

Considerable developments in concrete finishers have taken place in recent years. Initially, the finish as produced by a vibrating beam was considered sufficient as a running surface, but rising standards soon made it necessary to use a transverse reciprocating screed to remove irregularities. These screeds were originally hung from the rear of the finishing machines so that they rested on and were guided by the tops

FIG. 33. A concrete urban motorway under construction by fully mechanised methods. Bar reinforcement supported on stools is being used.

Fig. 34. A longitudinal joint being formed in the plastic concrete by a device for inserting preformed jointing material.

of the forms. A considerable improvement occurred when these screeds were hung centrally from a chassis because they were independent of minor irregularities in forms (Fig. 31). The application of the principle of articulation to finishing machines in conjunction with rigidly supported centrally hung screeds was a further step forward in producing high quality running surfaces. Variations on this theme have included angled screeds and longitudinal screeds. A final smoothing and levelling of the surface is sometimes carried out using a scraping straight edge made of flats or tubes; many contractors have their own ideas as to the most effective device to use. The final texture is most commonly formed by means of a steel broom dragged across the surface from a simple mobile bridge, but even this process can be done by machine.

*Concrete Roads* 71

FIG. 35. A contraction joint formed in plastic concrete and sealed with a removable rubber strip. Permanent sealing is carried out later. Note that the contraction joint is already functioning.

It is common practice on well planned jobs to construct a trial length of the same thickness of slab as that to be laid in the work to the full width of the finishing machine and of a length of at least 200 ft in order to act as a trial for the concrete mix and for the plant to be used. Construction of this trial slab should be commenced sufficiently in advance of the main construction to enable cores to be cut from it and examined. The trial slab should be repeated if the necessary standards of riding quality, concrete quality and joint construction are not reached. Figure 32 shows a complete train of concrete plant laying a

Fig. 36.  Fig. 37.

Fig. 36. A prefabricated contraction joint assembly in position ready for concreting. The assembly is nailed to the sub-base to prevent movement.

Fig. 37. Temporary sealing of the top of an expansion joint with paper rope.

two-lane carriageway. Figure 33 shows an urban motorway under construction by fully mechanised methods.

JOINTS

Expansion joints, contraction joints, longitudinal joints and construction joints are all used in major projects, although construction joints

Concrete Roads 73

FIG. 38. A power operated brush for cleaning out joints prior to sealing.

are used only rarely as every effort is made to cease operations at the position of an expansion or contraction joint. Sawing (see Fig. 21) is used for forming joints, or preformed plastic or composition strips are placed in the plastic concrete to form a sealing groove (see Figs. 34 and 35). Joints should be sawn as soon as is possible without excessive spalling if random shrinkage and temperature cracks are to be avoided. In heavy duty roads all transverse joints are usually provided with dowel bars and these are held in place in prefabricated cradles during concreting (see Fig. 36). The sealing of joints is often delayed until the concrete has dried out, and during the interim period grit can enter joint grooves and cause spalling of the arrises. Figure 37 shows one method of preventing damage of this nature by using a temporary filling.

A variety of materials has been used for filling joints in an attempt to provide a material which will have good adhesion to concrete, which will have the required elasticity and which will be stiff enough at all times to resist the ingress of foreign matter. These demands are very exacting and no perfect material has yet been discovered. For this reason, closer joint spacing which results in less opening at individual joints has been encouraged by the shortcomings of existing joint sealing compounds. Most joint filling compounds require that the concrete at joints should be cleaned and dry before being placed in the joints if adhesion is to be satisfactory. Joints should be slightly undersealed to prevent unsightly distribution of excess material over the surface in hot weather. To reduce tensile stresses in joint sealing compounds, the width of the sealing material when placed should be as large as the depth. To achieve this object, deep sealing grooves should be caulked with an inert material like bituminised hemp. Figure 38 shows a power operated brush for cleaning out joints prior to sealing, whilst Fig. 39 depicts a joint sealing machine.

CURING

It is usual to protect the finished surface of a concrete slab for 2 or 3 hours so that it will not be marked in the event of heavy rain. In addition, the concrete is cured either by covering with materials such as plastic sheeting, hessian or waterproof paper or more commonly by spraying with a resinous curing compound. A white pigmented compound is desirable as it reduces heat loss and absorption. These resinous curing compounds weather after a few weeks when they are no longer required and are soon removed from the surface. A mechanical sprayer is shown in Fig. 40.

---

FIG. 39 (*opposite*). A machine for melting and pouring hot joint sealing compound into joints. The rate of flow is readily controlled by the operator.

Fig. 40. A self-propelled automatic membrane curing compound sprayer.

OPENING THE ROAD TO TRAFFIC

The guide previously given for roads constructed by semi-manual methods applies to the time which should elapse before slabs can safely carry traffic.

MEASURING RIDING QUALITY

The difficulty about obtaining good riding quality on a concrete road slab is that it is not possible to do much in the way of remedial measures after the concrete has hardened. Grinding of high spots can be done, but the areas which can economically be treated in this way are limited. It is highly desirable therefore to expend as much effort as is necessary to ensure accurate surfaces on the plastic concrete. A wet surface profilometer has been developed by the Road Research Laboratory as a research tool to give an autographic record of the quality of the running surface at two, four or more points as desired, across the width of the slab. It does, however, suffer from the defect that the record it produces is the irregularity of the slab in relation to the forms upon which the apparatus runs.

A further apparatus, also developed by the Road Research Laboratory, is the multi-wheeled profilometer[11] which sums inches of vertical irregularity, the result being quoted per mile of length as "q" figures (Fig. 41). Figures between 40 and 50 in. per mile give excellent riding quality and indices as low as 20 have been achieved in concrete.[12]

## Slip Form Paver

The slip form paver is a type of concrete road laying machine originating in the United States of America[13] and able to lay concrete without fixed side forms. The machine embodies all the functions of a spreader and finisher in one and spreads, compacts and forms the concrete into a pavement by extrusion through an orifice while temporarily confining it between trailing side forms. A variety of these machines exists, the simplest requiring prepared strips upon which the crawler tracks run in order to provide line and level of the finished surface. In its most highly developed form the slip form paver is guided by a sensing device which feels its way along stretched piano wires fixed to correct line and level on both sides of the machines (see Fig. 42).

Slip form pavers were first developed in America about 1949 for major road contracts, but now smaller editions are available and under development for use in the construction of minor concrete roads. One example is shown in Fig. 43.

In all types the mode of operation is essentially the same.[14] The

FIG. 41. A multi-wheeled profilometer used for measuring the riding quality of surfaces.

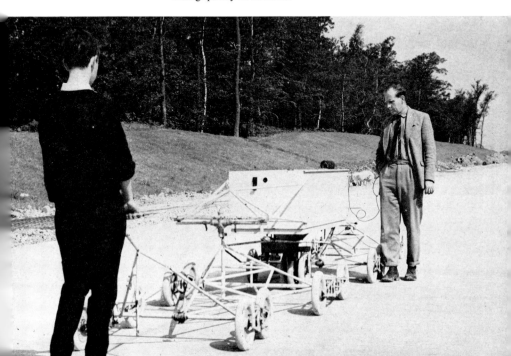

FIG. 42. The sensing device on a wire-guided slip form paver.

loose concrete is spread in front of the machine or dumped into a hopper forming part of the front of the paver (Fig. 44) and is then struck off approximately to the correct surcharge, which is, however, not critical. The concrete is then vibrated by a series of poker vibrators or by a vibrating beam buried in the concrete, which drives out the air and renders the concrete in a fluid condition. This fluidity is important and is achieved by intense vibration. The forward motion of the paver absorbs the concrete whilst it is in this fluid condition and extrudes it through a shaped orifice, thus forming the slab. By the time the concrete emerges it is beyond the effective range of vibration and it retains its extruded shape. The workability of the concrete mix is somewhat surprisingly not critical within the limits of 1 to 3 in. slump, but clearly if the fluidity is too high the formed edges will slump sideways. Too low or too high a sand content can also cause edge slump. Reinforcement is clearly more difficult to introduce into a slab formed by a slip form paver, but a variety of methods have been tried successfully.[15] Dowel bars at transverse joints and tie bars in longitudinal joints in multi-lane construction can also be incorporated. Little final surface finishing is required except texturing, which can be done either by brushing or by burlap drag, i.e. a strip of hessian is dragged along the finished surface. A scraping straight edge or long-handled float is often used to improve small areas which may be open in texture. The output of slip form pavers can be as much as 400 linear feet per hour, which is much higher than is usually achieved with conventional machinery. Attainment of this large rate of output is of course dependent upon having the large quantities of materials available, coupled with large capacity batching and mixing plant.

Slip form pavers were originally developed in the road field for producing concrete sub-bases but as the quality of the surface finish has improved with experience, they have, in recent years, also been used for producing concrete running surfaces. The potential of slip form pavers in this respect is high because the riding quality of the pavement produced appears to be largely independent of variations in the loose density of the spread concrete and because of the design of the machines, which is basically that of a tracked vehicle. Figure 45 shows a completed slab.

Heavy rain produces difficulties in slip form paver work in eroding the edges and causing slumping of the edge of the pavement. The former difficulty is, of course, no greater than occurs with conventional machinery, whilst the latter has been controlled in practice by propping the lower edges of the pavement with planks of wood until the concrete has stiffened. One advantage arising from the very large output of the slip form paver is that work can be confined to periods of fine weather.

FIG. 43. A British slip form paver.

Fig. 44. Concrete tipped into the front of a slip form paver.

Slip form pavers can be used to construct a slab adjacent to the existing one by running one track in a raised position on the edge already laid.

The use of slip form pavers has resulted in some saving of cost,[16] which arises because of the elimination of fixed side forms and of the relatively large number of units used in conventional work. The

Fig. 45. A slip form paver laying a 10 in. thick slab 24 ft wide with levels taken from a stretched wire.

Concrete Roads    81

machine is mobile about the site under its own power, which means that it is much more flexible than a concrete train. A further advantage is that slip form paver work does not require the relatively highly trained team of men which conventional machinery demands. Finally, the riding quality of pavements produced with a slip form paver is likely to be better than that produced hitherto.

### References

1. ROAD RESEARCH LABORATORY. *Annual Report for 1961*, London, H.M.S.O., pp. 134.
2. BLAKE, L. S. and BROOK, K. M. *The construction of major concrete roads in Great Britain—1955–1960*, Technical Report TRA 353, London, Cement and Concrete Association, June 1962, pp. 43.
3. ROAD RESEARCH LABORATORY. *Protection of subgrade and granular bases by surface dressing*. Road Note No. 17, London, H.M.S.O., 1953, pp. 4.
4. CEMENT AND CONCRETE ASSOCIATION. *Concrete roads: 3—Design of mixes*, Series Db. 9, London, 1955, pp. 16.
5. SHACKLOCK, B. W. Air entrained concrete—properties, mix design and gravity control, *The Surveyor*, 27 August 1960.
6. CEMENT AND CONCRETE ASSOCIATION. *Making test specimens on the site*, Series Bm 10, London, 1958, pp. 14.
7. BRITISH STANDARDS INSTITUTION. B.S. 1926, *Specification for ready-mixed concrete*, 1962, London, pp. 11.
8. SHARP, D. R. The scraping straightedge—a simple tool for improving the riding qualities of concrete surfaces, *The Surveyor*, vol. 118, No. 3512, 26 September 1959.
9. CEMENT AND CONCRETE ASSOCIATION. *Housing estate roads*, Series Db. 14, London, July 1961, pp. 33.
10. CEMENT AND CONCRETE ASSOCIATION. *Bay layout for concrete roads*, Series Db 24, London, 1964, pp. 17.
11. SCOTT, W. J. O. *Roads and their varying qualities*, Road Paper No. 25, Institution of Civil Engineers, London, 1948.
12. KIRKHAM, R. H. H. *The riding quality of concrete roads*, Road Research Technical Paper No. 60, London, H.M.S.O., 1963, pp. 87.
13. SHARP, D. R. America—a generation in advance. London, *Highways & Bridges*, 28 August 1963.
14. BURKS, A. E. Application of the slip-form paver to British road construction, *Cement, Lime and Gravel*, vol. 40 (5), July 1965, pp. 173–178.
15. WALKER, B. J. *Concrete pavement achievements in Great Britain*. Fifth World Meeting of the International Road Federation, London, September 1966. Cement and Concrete Association, pp. 8.
16. BURKS, A. E. and MAGGS, M. F. The Cromwell slip form paver trials, *Proceedings of the Institution of Civil Engineers*, vol. 36, February 1967, pp. 225–271.

# CHAPTER 5

# Cement Stabilised Bases

**Introduction**

The stabilisation of natural soils and aggregates is now a widely used form of construction for road bases. Traditional base construction methods using mechanically stable material are not always suitable for present-day heavy traffic intensities and loadings, but they can readily be improved by the incorporation of cement. The use of cement, moreover, permits inferior "waste" materials to be used, with subsequent economy. A number of cement stabilised or cemented materials are commonly recognised according to the aggregate used. Table 3

TABLE 3. ROAD CONSTRUCTION MATERIALS CONTAINING CEMENT

| Material | Part of road | Raw material | Testing (preliminary and control) | Mixers used |
|---|---|---|---|---|
| Concrete | Riding surface | Processed aggregate | Some essential | Free fall |
|  | Base | Processed aggregate | Little | Free fall |
| Lean concrete | Base or sub-base | Processed aggregate | Little | Free fall |
| Cement bound granular material | Base or sub-base | Selected from fairly wide range of granular materials | Some essential | Usually forced mixing |
| Soil-cement | Base or sub-base | Selected from wide range of soils | May be considerable | Forced mixing |

shows their similarities and differences. To make the table complete, data relating to concrete have been included.

Soil-cement has been used in the United States since 1935 and in Britain since about 1948. The use of lean concrete on a large scale commenced in Britain in about 1954 as a development from soil-cement, being a material more suited to roads carrying heavy traffic and also (paradoxically) for work where control is of a low standard. Cement bound granular material is a further development from lean concrete whereby some relaxation of grading of the constituent aggregate is permitted.

**Soil-cement**

Soil-cement is a compacted mixture of pulverised natural soil and cement mixed with the appropriate amount of water. The resultant material has the appearance of very dense soil, the colour being almost the same as that of the soil used in construction. It is hard, does not readily dust, but has a limited resistance to abrasion.

DEVELOPMENT AND APPLICATIONS

The idea of using cement to strengthen soils is not new, and indeed a British engineer—H. E. Brooke-Bradley—combined soil with cement on Salisbury Plain in 1917 to impart stability to muddy tracks to enable them to carry traffic. During the 1939 war, soil-cement was used in Britain and more extensively in Germany for airfields; its use for roads in Britain commenced in about 1948. Soil-cement has now been used widely in Britain for such diverse purposes as car parks, cycle tracks, footpaths, parade grounds and foundations for structures such as tanks and for building.[1] For concrete roads and airfield runways, soil-cement provides a base with a smooth surface on which to superimpose the concrete slab. It also provides a weatherproof means of access for contractors' plant, materials and equipment. Although it is not strictly necessary for good performance, any extra cost of providing a stabilised layer under a concrete slab is usually recovered because of the ease with which construction proceeds.

It is clear that soil-cement work in Britain has not enjoyed the popularity which it deserves. This is probably due to the fact that soil-cement

## Concrete in Highway Engineering

techniques require knowledge which is not part of the equipment of the average engineer, who must also be prepared to exercise rather greater control than he would do with most types of highway construction. Contractors too often consider quite erroneously that soil-cement construction is "difficult" and weather sensitive—in fact, the process can be carried on except in the worst of winter conditions. Without a doubt, soil-cement is by far the cheapest form of road construction and for this reason alone it merits much greater attention. Economy arises primarily from the fact that the majority of the raw material needed—the soil—is taken from or near the site and costs virtually nothing.

### CHOICE OF SOILS

Most soils can be stabilised with cement although it may not be economic to use the high percentages required in some cases. Exceptions are highly organic soils such as agricultural top soils and peat. Heavy cohesive soils present difficulties in breaking them up sufficiently to mix in the cement. In some cases granular material has been added to a pulverised cohesive soil in order to reduce its cohesive properties. Any non-porous, waste, granular material can be used, e.g. slag dust, sand.

### SOIL SURVEY

The first step in a soil stabilisation project is to ensure that the soil will respond to treatment and for this purpose it is necessary to carry out a soil survey to provide information on soil conditions over the site, including the level of the water table. The soils are located, identified and sampled at such frequencies as to be truly representative of the soils to be processed.[2, 3]

### SOIL TESTING

The following stages are involved in soil testing:[4]

1. Soil suitability tests—preliminary testing to establish whether soils are suitable for stabilisation with cement.
2. Main test—laboratory testing carried out on the principal soil types to provide information for design and specification purposes.

## Cement Stabilised Bases 85

3. Uniformity test—laboratory testing on a larger number of small soil samples to check the validity of the main test results.
4. Site test to check the quality of the work.

Clays with a liquid limit up to 40 to 45% are generally regarded as being possible to stabilise. With granular soils the following is a guide for selecting those suitable for processing:

| | |
|---|---|
| Maximum size | 3 in. |
| Passing $\frac{3}{16}$ in. B.S. sieve | above 50% by weight |
| Passing No. 36 B.S. sieve | above 15% by weight |
| Passing No. 200 B.S. sieve | below 50% by weight |
| Finer than 0·002 mm | below 30% by weight |

MOISTURE CONTENT

The effect of moisture content on the quality of soil-cement arises largely from its influence on the state of compaction obtained in the field. For high densities—and these are essential for durability and strength—it is necessary to bring the materials being processed to the best moisture content for compaction with the equipment available. For this reason, standard dry density and moisture content tests, in accordance with B.S. 1924,[3] are used to decide the optimum moisture content. The moisture required for the hydration of the cement is adequately provided by the moisture necessary for maximum compaction. This is confirmed by tests which show that the greatest strength is obtained at approximately the optimum moisture content for compaction.[5] As a guide in the case of plastic soils, the optimum moisture content from considerations of density and practical expediency approximates to the plastic limit. In the case of sandy soils, the British Standard compaction test is not reliable and it is more satisfactory to decide the moisture content from a cube test as described later in this chapter for lean concrete.

CEMENT CONTENT AND DURABILITY

The proportion of cement required is a most important item of knowledge and the test consists of making and testing samples with different cement contents to determine which is most satisfactory for strength with economy. The unconfined compressive strength test described in B.S. 1924[3] is used in Britain for fine and medium grained soils, using cylinders with a height : diameter ratio of 2. The diameter of the specimen is 2, 4 or 6 in., depending upon the particle size distribution of the soil. The test specimens are tested in compression, usually at 7 days after curing by coating with paraffin wax. A compressive strength of 250 lb/in$^2$ at 7 days is usually accepted as the desirable minimum in Great Britain. The original derivation of this figure is obscure, and although it has stood the test of time there is no particular merit in it as such. A figure of 400 lb/in$^2$ is probably more appropriate. For medium and coarse grained soils, 6 in. cubes are used.[3]

The cement content affects not only the strength and therefore the load bearing capacity of soil-cement, but also the durability. In practice, a road base must be able to withstand the effects of water and frost as well as applied loads. Standard tests to examine the effects of water and freezing and thawing on soil-cement have been devised but are not widely used in Britain because of the difficulties of interpreting the results and relating them to practical conditions.

It has been found in practice that soils such as silt and chalk, which in their natural state exhibit great susceptibility to frost heave, can be satisfactorily stabilised and rendered immune to frost action by mixing with cement. The effects of sulphates contained in soils or soil-cement are examined by means of a B.S. 1924 test in which the 14-day compressive strength of cylinders cured normally is compared with the strength of similar cylinders after soaking in water.

Some organic compounds in soils inhibit the hardening action of Portland cement and B.S. 1924 contains a test to detect the presence of these compounds based on the pH of the soil-cement paste. The use of extra-rapid hardening Portland cement, which contains about 2% of calcium chloride, often neutralises the organic compounds and the soil-cement then hardens at the normal rate.

CONSTRUCTION PROCEDURES

The basic operations in the construction of a soil-cement base are as follows:

1. Pulverisation of the soil.
2. Intimate mixing of the soil with cement and water.

## Cement Stabilised Bases

3. Compaction of the mixture at the optimum moisture content to the highest density possible.
4. Curing.

There are three methods of construction, as follows:

1. Mix-in-place methods: (a) by multi-pass plant, (b) by single-pass plant.
2. Premix or stationary plant method.
3. Travelling plant mix method.

The mix-in-place method is most applicable to cases when the soil on the site of the construction is being used. The premix method is used when the soil has to be imported, but even in this case, imported soils are sometimes spread over the area and processing done by mix-in-place methods. The travelling plant method involves the use of large, expensive equipment and its use will be economic only when large areas of construction are involved. It has been used in Britain on motorway work.

### 1. MIX-IN-PLACE METHODS

In all mix-in-place operations, initial preparation of the ground is important. After removing all top soil the ground should be shaped up to the finished levels of the soil-cement base to an accuracy of not less than $\pm\frac{1}{2}$ in. under a 10 ft straight edge.

It is also important to ensure that the initial density of the soil is uniform throughout the area to be processed, since uneven initial density will result in unevenness in the finished soil-cement surface. It is very difficult and certainly undesirable to attempt to reshape the compacted soil-cement slab even before it is hardened, and care in the initial preparation of the soil is amply repaid.

(a) *Multi-pass method.* The soil is first pulverised to break it down sufficiently to ensure intimate mixing with water and cement. Typical equipment used for the purpose is shown in Fig. 46. The depth to which the soil is pulverised should be such as to give the required finished thickness after compaction. On completion of pulverisation, the soil is compacted lightly and graded to the required crossfall or camber. If water is to be added to the soil it should in the case of

FIG. 46. A 4-ft-wide offset multi-pass mixing machine fitted to a standard tractor.

granular soils be incorporated at this stage. With more cohesive soils, it is better to delay the addition of water until after the soil has been mixed with the cement. The water is distributed using a machine with a spray bar, preferably offset, i.e. the water tank runs at the side of the area being processed. After adding water it is mixed thoroughly and uniformly to the full depth of the loose soil.

Cement can be spread mechanically or by hand. The mixing of the soil with cement is usually done with the same plant as that used for pulverisation. Mixing is continued until the soil-cement is of uniform colour and texture over the full area and to the full depth of treatment. The moisture content of the mix is again tested and more water added if required. After final mixing the shape of the surface is restored and compaction carried out. Multi-wheeled pneumatic rollers, smooth-wheeled rollers, vibrating rollers and power rammers are all used for compacting soil-cement, the latter equipment being particularly suitable for small areas. Concrete road compacting and finishing equipment has also been used on occasions, particularly for granular soils.

After compaction the surface of the soil-cement must be prevented from drying out too quickly, and curing is essential. An immediate application of bituminous emulsion at a rate of about 8 yd$^2$/gall is most commonly used, followed by sanding if the base is to be trafficked to any extent before surfacing. Traffic is kept off the processed soil-cement for 7 days to allow for an appropriate gain in strength.

(b) *Single-pass method.* The single-pass method of construction depends upon the use of equipment specially designed for the purpose and British, German and American plant is at present available. As the name implies, the complete operation of constructing the soil-cement pavement is completed in one pass of the equipment, which takes various forms. Some machinery embodies the whole of the pulverising, mixing, wetting, cement spreading and compacting equipment in the one machine, whilst in others separate machines do one or more of the operations, although the whole is formed into a single train. Typical British equipment complete with dropping weight compactor is shown in Figs. 47 and 48. Figure 49 shows a large output American machine with four rotors under the mixing hood.

In the single-pass method the first operation is to remove the turf and top soil and, as with the multi-pass method, the surface is corrected to the crossfall and levels required for the finished road. The machine then moves forward at a speed varying from about 3 to 25 ft/min, according to the power available and to the type of soil being processed. A fully completed soil-cement pavement emerges from the rear of the train, the only operation remaining to be done being the removal of minor irregularities in the surface with a smooth-wheeled roller. An advantage of the single-pass method is that the depth of processing can be more accurately controlled, since the relatively uniform level of the undisturbed soil is used as a datum. A further advantage is that the risk of damage by rain is reduced to a minimum as only sufficient cement is spread as is required for immediate use in the process. The fact that the soil-cement is compacted so soon after mixing ensures that the amount of freshly mixed material exposed to inclement weather is virtually nil. Finally, the risk of low densities due to delay in compaction —a greater possibility with cohesive soils than with granular material—

FIG. 47. A cement spreader (background) and single-pass machine mixing cement into soil to produce a soil-cement sub-base for a motorway.

Fig. 48. Single-pass soil-cement equipment on airfield construction. Note the bagged cement supply to the cement spreader incorporated in the front machine with the water spray system fed from the bowser. The rear machine is a dropping weight compactor.

is non-existent. After the passage of the machinery the remainder of the work—curing and control testing—is the same as for the multi-pass method.

2. PREMIX OR STATIONARY PLANT METHOD

In the premix or stationary plant process, the soil and cement are mixed together in suitable mixers, and the material thus produced is carried to the site, where it is spread and compacted. An advantage of the method is that it permits good control of the process of mixing and of the moisture content and it is easy to ensure a uniform thickness of processed material in the pavement. The disadvantages are that it is

Fig. 49. A large output single-pass stabilising machine constructing a sub-base for a concrete road. Cement is spread on the formation in advance of the machine and water is supplied by a linked bowser.

usually a slower form of construction than the mix-in-place method, requires more plant and labour, and is usually more expensive. It is particularly suitable where the natural *in situ* soil is not suitable for stabilisation and it is therefore necessary to use an imported material. This material can be obtained from a borrow pit, quarry or from another part of the construction site.

In the premix method, the sequence of operations is as follows:

(a) Site preparation.
(b) Excavation or importation of soil to be processed.
(c) Mixing.
(d) Transporting.
(e) Spreading.
(f) Compaction.
(g) Curing.
(h) Surfacing.

To prepare the site for premixed soil-cement work, the turf and top soil are removed and the formation prepared as described for the mix-in-place method, except that the level of the formation must allow for superimposing the soil-cement and be thoroughly compacted beforehand. The soil, having been tested and found suitable, is excavated and stockpiled near the mixing plant, which may be located at the source of supply or adjacent to the road construction site. Ordinary concrete mixers, both tilting and non-tilting types, and continuous mixers are satisfactory for the more granular soils, but with the more cohesive types, the best results are obtained with pan type mixers, pug mills and paddle mixers. In these, lumps of soil are more easily broken up and the cement intimately mixed in. Mixing is continued until the soil-cement is of uniform colour and texture. Batching is preferably done by weight.

The shortest possible time should elapse between mixing and spreading and compacting the soil-cement. It is desirable that the compaction process should be complete within 1 hour of mixing and if the material is exposed to the weather in the transporting vehicles, it should be protected by tarpaulins in warm dry periods or in rain.

The thickness of the material which can be spread and compacted in one layer is limited because of the difficulty in obtaining high densities

Fig. 50. Laying a plant mixed cement stabilised sub-base with a bituminous finisher. Compaction is by tandem roller.

throughout considerable thicknesses. With modern compacting equipment as much as 9 in. or more of material can be compacted in one layer, but the technique chosen for any particular case should be checked by tests on the density being obtained. Spreading can be carried out by hand, by bituminous spreader (see Fig. 50) or by concrete spreading machines. Simple blade spreaders do not give high quality work. Compaction is as for the mix-in-place method.

3. TRAVELLING PLANT METHOD

In the travelling plant method, the pulverised soil is drawn into a windrow by a motor grader or other suitable plant and the cement spread on top. After mixing, the material is discharged on to the road formation and spread, ready for final compaction. The travelling mixer moves along the line of the windrow, lifting the soil and cement into it. The machines available for this process are mainly American and are expensive, although they have very large outputs. They have not been widely used in this country.

**Lean Concrete**

Lean concrete is a mixture of aggregates complying with the requirements of B.S. 882, *Concrete Aggregates from Natural*

*Sources* and cement mixed and compacted with the appropriate amount of water. It is, as the name implies, a lean concrete because the amount of cement used—about 5%—is much less than is contained in the traditional concrete slab. It can be mixed in free-fall concrete mixers and since it is made from "standard" materials the need for preliminary testing is eliminated and the amount of site control required is reduced. These factors make lean concrete particularly useful on smaller jobs

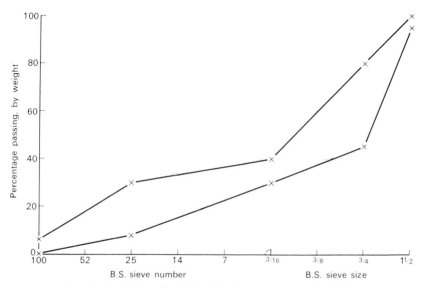

FIG. 51. Typical grading limits for lean concrete using 1½ in. maximum size aggregate.

where the extra cost of using concreting aggregates will be offset by the convenience of being able to use standard concrete mixers and by the reduction in control testing and supervision required. The material has, however, been extensively used for all classes of roads in Britain, including motorways.

AGGREGATES

Any gravel or crushed rock aggregate complying with B.S. 882, *Concrete Aggregates from Natural Sources* can be used, but the requirements for hardness and silt content normally applied for concrete work can be lowered to some extent. The typical grading limits for $1\frac{1}{2}$ in. maximum size crushed rock or gravel aggregates are shown in Fig. 51. A continuous combined grading of coarse and fine materials should be aimed at, but if this cannot be achieved economically, it is important at least to ensure that there is no deficiency of fine material. Experience has shown that the proportion of sand, i.e. material passing the $\frac{3}{16}$ in. British Standard sieve, should be between 30% and 45% by weight of combined aggregate. The most suitable proportion of sand can be tested by trial at the start of the work and is generally greater for coarse sand than for fine sand. It is, however, recommended that the proportion of material passing the No. 25 British Standard sieve should lie between 12% and 25% by weight of combined aggregate. If the mix tends to segregate or if it is found that the surface of the base is not sealed on completion of compaction, the proportion of fine material should be increased by adding sand. All unwashed gravel aggregates and also aggregates not strictly complying with the requirements for silt content given in B.S. 882 should be examined for stability by making a trial mix with a cement content of 6% by weight of dry materials. If cubes compacted to refusal as described later and tested in compression at 28 days give compression strengths in excess of 1500 lb/in$^2$ (1000 lb/in$^2$ at 7 days) the aggregate can be accepted for use in lean concrete. If fully compacted cubes made from this mix fail to reach this strength, the cement content may be increased to 230 lb/yd$^3$ (15 : 1), but any failures at this strength of mix should result in rejection of the aggregate.

### CEMENT

Ordinary Portland or Portland blast-furnace cements are chiefly used in lean concrete, but rapid hardening or extra-rapid hardening cements may be used if the base must carry traffic as early as possible.

### MIX DESIGN

Contrary to conventional practice with ordinary concrete, the water content of lean concrete for compaction by rolling is decided solely on the amount required to give optimum compaction and the approach, therefore, is similar to that for soil-cement. Compaction trials[5] and

Cement Stabilised Bases 95

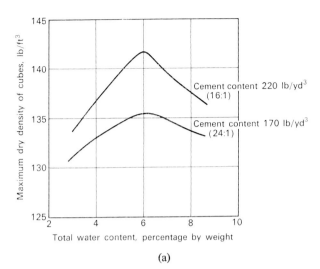

(a)

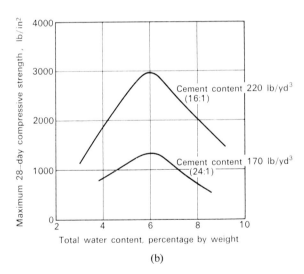

(b)

FIG. 52. Typical curves showing the influence of mix water content on (a) the maximum dry density and (b) the strength obtainable from lean concrete cubes compacted to refusal. These values apply to an irregular Thames Valley gravel aggregate.

## 96 Concrete in Highway Engineering

site experience[6] have shown that for compaction by rolling, the optimum moisture content (i.e. the total water in the mix including water absorbed by the aggregate) is about 6% by weight of the dry materials. Mixes varying in cement content between 550 lb/yd$^3$ (6 : 1) and 140 lb/yd$^3$ (24 : 1) have been compacted successfully using this value of moisture content, although somewhat higher figures have been required for highly absorptive aggregates and for aggregates containing high proportions of silt. The water content should, in practice, be as high as possible, consistent with obtaining a level and sealed surface by rolling. If the mix is too dry, the surface of the lean concrete may shear under the roller and appear loose after compaction. If the mix is too wet, concrete will be picked up on the wheels of the roller, which will tend to sink into the concrete. Figure 52 shows typical compaction curves obtained when using irregular Thames Valley gravel aggregate, and indicating the influence of moisture content on the dry density and compressive strengths of cubes compacted to refusal.

CEMENT CONTENT

The earlier lean concrete bases used about 250 lb of cement per cubic yard (14 : 1), but more recently engineers have tended to favour leaner mixes. Experience has shown that for all-round results it is probably best to use a cement content of between 170 and 230 lb/yd$^3$ (20 : 1 to 15 : 1).

COMPRESSIVE STRENGTH AND WATER : CEMENT RATIO

Although the compressive strengths of lean concrete are related to the water : cement ratio in much the same way as richer and more workable concrete, this approach need not be used when designing a lean concrete base, because, as has already been pointed out, the water content required for satisfactory compaction by rolling is constant and is practically independent of cement content. It has become current practice on more important projects to specify the minimum strength to be achieved.[7] Strengths are measured by crushing lean concrete cubes made as described in B.S. 1881, except that considerably more effort is required to compact these mixes than is implied in the British Standard. The most satisfactory method is to apply an electric or pneumatic hammer to the material, which is filled into the mould in three layers. In practice, the lean concrete base cannot always be compacted by rolling to such a high dry density as a cube compacted to refusal. Laboratory tests show that a reduction in dry density of 5% causes a reduction in compressive strength of about 40%. With good site compaction, the dry density of the pavement can be between 95% and 97% of cubes compacted to refusal. It follows that the compressive strength of material in the base can be between 60% and 75% of that measured in cubes.

CONSTRUCTION

When the aggregate grading limits shown in Fig. 51 are adopted, lean concrete can be produced by any of the batching and mixing plants commonly used for ordinary concrete. On small jobs, e.g. road widenings, concrete can be produced in a mixer as small as 10 ft$^3$ capacity or ready mixed concrete can be used. On larger jobs, it is usually desirable to use mixing plants with capacities of more than 20 yd$^3$ per day. It is best to have aggregates supplied in two separate sizes rather than "all-in" because this permits better control over the proportion of fine aggregate in the mix.

Fig. 53. Laying lean concrete using a bituminous paver to form the base for a motorway.

Fig. 54. A paver laying lean concrete. Note the rear mounted compacting beam with vibrators.

The mixed material is usually transported in lorries with covers available to protect the mix from wetting by rain or evaporation during dry weather. The length of haul must be limited by the requirement that compaction be completed within $1\frac{1}{2}$ hours of the water being added to the cement. Lean concrete can be spread easily by shovel or on larger jobs by mechanical equipment. The type of equipment used has included graders, dozers, bituminous finishers and concrete spreaders (see Figs. 53, 54 and 55). Although it is not essential to use forms, it is usually advantageous, but alternatively kerbs and/or channels may be placed before concreting, in order to assist in controlling the levels of the spread and compacted concrete. A scratch templet operating from the forms, kerbs or channels is of great assistance in checking levels. It is important that the lean concrete should be spread to an adequate surcharge to allow for compaction. The amount usually is about 25% of the compacted thickness but trials should be carried out on the site. When spreading lean concrete, particular care must be taken to avoid longitudinal joints, which are a source of weakness, and it is therefore better to spread the concrete full width over the carriageway. When using a mechanical spreader, the length of lane should be limited so that a free longitudinal edge of spread material is not left for more than 1 hour.

Where the thickness of lean concrete required exceeds 8 in. it is doubtful whether adequate compaction can be obtained throughout the full depth of the slab. Two-layer spreading and compaction is essential in this case unless trials show that it is unnecessary. The top layer should be laid immediately after compacting the lower layer so that in any section the material is fully compacted throughout its full depth within $1\frac{1}{2}$ hours of mixing the first batch in that section. Alternatively, placing of the top layer must be delayed until such time as the lower layer can safely carry the essential construction traffic. In normal weather, the delay need not be more than 3 days.

Either smooth-wheeled or vibrating rollers or rubber-tyred rollers may be used to compact lean concrete. Six passes of a 6–8 ton smooth-wheeled roller are usually sufficient to produce satisfactory compaction, but an initial two passes made with a lighter roller are advantageous to bed down the material so that it will not push in front of the heavier roller. Vibrating rollers are preferable when the output of lean concrete is low because they can be used efficiently over the smaller areas. In using vibrating rollers, eight passes are usually sufficient to give good compaction, the first two passes being without vibration to bed down the material. Particular attention should be paid to compaction at joints, which should all be vertical. Hand or mechanical rammers are useful here.

Fig. 55. Laying lean concrete with about 12% moisture content with a slip form paver.

100     *Concrete in Highway Engineering*

CURING AND OPENING TO TRAFFIC

Lean concrete bases require curing to prevent evaporation of moisture. The most convenient method is by spraying a bituminous emulsion or cut back bitumen onto the surface at the rate of 1 gall. to 6–8 yd$^2$ (see Fig. 56). This layer can be sanded or chipped if it has to carry traffic.

Fig. 56. Bituminous emulsion being sprayed on a lean concrete base for curing.

When there is a delay in laying the top course of lean concrete in two-course work other methods of curing should be adopted for lower layers, such as spraying water over the surface at least twice a day. No traffic should be allowed on a lean concrete base until at least 3 days after laying, and longer in cold weather. If the surfacing is not laid as soon as the 3-day period has elapsed, traffic should be kept to a minimum until the base is 7 days old. If at all possible, the base course surfacing should be laid before the road is open to full traffic, but if this is not possible a double surface dressing should be maintained. In case of urgent need, lean concrete bases can be laid, surfaced and opened to traffic all in the one day without much risk of failure.

## Cement Bound Granular Material

Cement bound granular material is similar to lean concrete except that the grading limits of the material used are rather wider. Suggested limits are given in Table 4.

TABLE 4

| British Standard sieve size | Grading range % |
|---|---|
| 2 in. | 100 |
| $1\frac{1}{2}$ in. | 95–100 |
| $\frac{3}{4}$ in. | 45–100 |
| $\frac{3}{8}$ in. | 35–100 |
| $\frac{3}{16}$ in. | 25–100 |
| No. 25 | 8–65 |
| No. 52 | 5–40 |
| No. 200 | 0–10 |

The advantage of cement bound granular material is that the aggregate is not so strictly controlled as that for lean concrete and it is consequently cheaper. The material can, therefore, be said to be between soil-cement and lean concrete as far as constituent raw material is concerned. This material will normally be as-dug gravel or hoggin and because of the greater likelihood of variability compared with washed, prepared and fully graded aggregates, it follows that more control testing of the material is necessary. This is no hardship on large jobs although the difficulty in practice is that if variations do occur, it is difficult to remedy them quickly. For this reason, the use of cement bound granular material has tended to decline, although for the smaller, less important jobs there is no reason why it should not continue to be widely used.

### Treatment at Joints

Most cement stabilised road bases have been laid without joints, except for day-work construction joints, and there is at present no evidence to show that expansion joints are necessary. The inclusion of these joints is clearly to be avoided as they encourage movement which will cause cracking in the surfacing. Construction joints at the end of the day's work should be vertical butt joints and special attention must be paid to thorough compaction of the material adjacent to these joints if compression failures are to be avoided in warm weather. Transverse day-work joints can be formed by feathering off at the end of the work, but it is most important to ensure that the material is cut back to a sound vertical face before work is resumed against it. If this is not done, expansion in the base may cause vertical differential displacement between one day's work and the next. A preferred method of forming day-work joints is to lay forms or heavy baulks of timber transversely, against which a roller can work. If necessary, there should be further compaction of the material by ramming to ensure high density.

### Surfacing with Bituminous Materials

Cement stabilised materials, whilst being capable of carrying heavy loads, do not have a surface which is resistant to traffic abrasion and it is therefore necessary to provide a permanent bituminous surfacing. A minimum requirement is a double surface dressing of tar or bitumen with applied chippings. In order to reduce further the risk of damage by frost and abrasion it is desirable to protect the surface by a bituminous carpet. This surfacing should be delayed until the base is at least 3 days old, or longer if cold weather has decreased the normal rate of gain in strength of the base. After 3 days have elapsed both the base and the wearing course can be placed. In cases of extreme urgency, however, the lean concrete base can be laid and surfaced and subsequently opened to traffic all within 24 hours without much risk of failure.

In choosing the bituminous surfacing it is important, as with any other form of construction, to use a wearing course of low permeability. A cement stabilised base will not permit drainage of water through it from a porous surfacing. The thickness of bituminous surfacing varies between 2 and 4 in. or more, depending upon the intensity of traffic.

It has been found that cracking in the base as a result of shrinkage and contraction of the cement stabilised material may sometimes cause cracks in the bituminous surfacing. Frequent narrow cracks are clearly preferable to widely spaced wide cracks as they will be less likely to affect the surfacing and will transfer load more efficiently. The exact cause of these differing crack patterns is unknown, although the nature and grading of the aggregate, the water content, the tensile strength of the stabilised material and the temperature and the time of year at which the base is placed are all likely to be relevant factors. These cracks, although they appear somewhat alarming, have not generally proved to be detrimental, although it is clear that if they are wide, water will penetrate through into the construction if they are not sealed. It is possible they will not be reproduced so markedly in the surfacing once it is replaced in the normal way of maintenance by new material.

**Compression Bumps**

Bumps have sometimes appeared in cement stabilised materials as a result of high compression forces. These occur more frequently in material placed in cool or cold weather and rarely after the surfacing has been applied. Areas of inadequate compaction as can be present adjacent to day-work joints encourage the appearance of these bumps, which are often confined to the top 4 in. or so of the base.

When these bumps occur the defective material should be replaced with fine bituminous material, or with new, fully compacted cement stabilised material if the damaged area is more extensive. It is unlikely that further trouble will be experienced.

### References

1. SHARP, D. R. Soil-cement as a structural and foundation material, *Proceedings of the Midland Soil Mechanics and Foundation Engineering Society*, vol. 2, paper No. 11, 1958, pp. 55–71.
2. ANDREWS, W. P. *Soil-cement roads*, London, Cement and Concrete Association, 1955, pp. 21.
3. BRITISH STANDARDS INSTITUTION. B.S. 1924, *Methods of test for stabilised soil*, London.
4. SHARP, D. R. Lean concrete and soil-cement in road and airfield bases, *Roads and Road Construction*, February and March 1960.

5. WILLIAMS, R. I. T. *A laboratory investigation of methods of compacting dry lean concrete test cubes*, Technical Report TRA 322, Cement and Concrete Association, 1961, London, pp. 18.
6. BLAKE, L. S. Lean concrete bases, *The Surveyor*, vol. 117, No. 3446, 10 May 1958, pp. 484–485.
7. MINISTRY OF TRANSPORT. *Specification for road and bridge works*, London, H.M.S.O., 1969, pp. 195.

CHAPTER 6

# Prestressed Concrete Roads

**Introduction**

In concrete roads, it is necessary to control cracking either by the use of reinforcement and/or by joints. An alternative is to arrange for the slabs to be in permanent compression so that the risk of cracks occurring is negligible. The application of an adequate amount of prestress is a convenient way of ensuring the condition of compression. Prestressing is beneficial, too, in a pavement that has cracked—for example, by overloading—because the compressive force keeps the cracks tightly closed. The first prestressed pavement was probably constructed at Nuzancy, France, in 1946. As early as 1955 Stott[1] listed twelve prestressed pavements which had been built in Great Britain.

**Advantages and Disadvantages of Prestressing**

The advantages of prestressing include the following:

1. By the application of a compressive force, crack opening can be prevented, resulting in a more durable pavement with better load carrying capacity.
2. Prestressed slabs can be made much longer than orthodox slabs and can even be constructed with no joints whatsoever, so simplifying construction of the slabs themselves.
3. It is likely from (2) that prestressed pavements will have a better riding quality than those constructed with joints.
4. There is the possibility that costs may be reduced as the slab thickness can be markedly reduced.

# Concrete in Highway Engineering

The disadvantages of prestressing include:

1. The repair of services beneath the road is difficult since the prestress will be disturbed and even lost when opening trenches in the slab.
2. The radius of vertical and horizontal curves which can safely be negotiated is limited, although experience[2] has shown that this is not a practical disadvantage because the curves that can safely be negotiated are more acute than those used on most roads (see Fig. 57).

FIG. 57. Prestressed concrete road constructed on curve.

3. Some knowledgeable engineer supervision is essential during the construction stage.
4. Higher standards of construction are necessary than for normal concrete pavements.

**Types of Prestressed Roads**

Prestressed roads may be classified into the individual slab type and the continuous type (see Fig. 58). The first consists of slabs separated

# Prestressed Concrete Roads 107

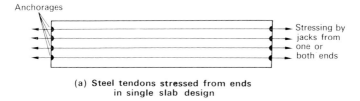

(a) Steel tendons stressed from ends in single slab design

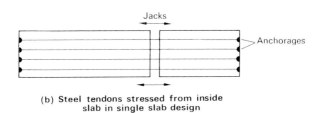

(b) Steel tendons stressed from inside slab in single slab design

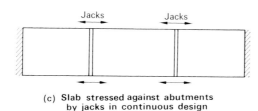

(c) Slab stressed against abutments by jacks in continuous design

FIG. 58. Methods of applying prestress to concrete slabs.

by expansion joints and prestressed by tendons, usually internal. Each slab is free to expand or contract to accommodate temperature and moisture changes in the concrete. Because the slab is free to move and is restrained only by the effects of subgrade friction, the amount of prestress in this type of slab will remain sensibly constant.

The second type consists of a continuous slab into which gaps to accommodate jacks are formed during construction. Abutments are constructed at each end of the slab to resist end thrust and after prestressing through the jacks, the gaps are blocked, so ensuring that the

108  *Concrete in Highway Engineering*

strain remains in the slab. As the slab ends cannot move, the stress in the concrete varies with temperature and moisture changes. As an alternative to jacks, the use of expanding cement has been used in America to induce compressive stresses in the concrete.

From the road users' point of view the continuous design has advantage over the individual slab type because the road appears as a continuous strip without any disfiguring joints or bumps. The continuous type, however, is more difficult to maintain in a satisfactory structural state because of the difficulty of ensuring that the amount of prestress is neither high nor low from a dangerous point of view.

**Individual Slab Type**

METHODS OF APPLYING PRESTRESS

Stress may be applied to individual slabs either by pretensioning or by post-tensioning.

(a) *Pretensioning*. In pretensioning the tendons are tensioned before the concrete is cast. After the concrete has hardened sufficiently the ends of the tendons are released, the stress in the tendons thus being transferred to the concrete through bond. In order to ensure that the prestress is uniformly distributed throughout the length of the concrete, the slab must be cut up into units depending upon the amount of subgrade friction. This method therefore suffers from the disadvantage that there will be present in the pavement free joints into which it is difficult to incorporate suitable load transfer arrangements. Moreover, these joints will vary in size depending upon the ambient temperatures and will function like expansion joints in a conventional concrete road. Rather surprisingly, the laying and stressing of the very long lengths of cable necessary in this method has not provided any great difficulty. On a trial length of motorway between Amsterdam and Rotterdam[3] the cables were stressed individually over a distance of about 1200 yd using a winch mounted on a lorry.

A runway at Vienna Airport has been constructed using this method, but here the tendons were stressed at relatively short intervals between abutments cast in the ground.

Prestressed Concrete Roads 109

(b) *Post-tensioning*. A number of the well-known systems have been used for post-tensioning concrete pavements together with some more ingenious adaptations. In the simplest systems, straight end anchored tendons are used. Other variations are perimeter cables which may be anchored anywhere along their length, diagonal tendons which are side anchored and longitudinal cables which are stressed by jacking at a gap left for the purpose in the slab. Strand and wire have both been used for tendons. Bond between the tendons and the concrete is prevented until the wires are tensioned and the ducts are afterwards usually filled with a cement grout. The ducts should be formed so that they will restrict the wires as little as possible in order to reduce the amount of prestress necessary. Ducts are usually formed by using a thin metal sheath which is sufficiently flexible to follow design curves, but rigid enough to keep its shape when the concrete is being compacted into position. Joints in the metal ducts must be made carefully to prevent the ingress of grout during the concreting operation.

Lengths of slabs stressed by post-tensioning have varied from quite short figures to over 900 ft. The length is governed primarily by the amount of subgrade friction, and the longer the slab the greater the need to induce the prestress as quickly as possible after placing the concrete to reduce the tendency for cracking due to shrinkage.

Measures adopted to minimise subgrade friction include careful preparation of the base and the use of slip layers and solid lubricants. Long slabs result in large movements at joints and Fig. 59 shows a design evolved especially for the purpose.

**Continuous Type**

In recent years the interest of highway engineers in the application of prestressing to roads has tended to become concentrated in the continuous type of construction. Here concrete is cast between abutments, gaps being left to accommodate jacks for prestressing. Freyssinet flat jacks in the form of thin sheet metal envelopes which can be inflated or screw and hydraulic jacks have all been used. Freyssinet jacks (Fig. 60) are sometimes inserted in series up to four or more in number, one or two jacks being used for the initial prestressing operation, the remainder being kept in reserve for subsequent prestressing

Fig. 59.    Fig. 60.

Fig. 59. Specially designed joint between prestressed concrete slabs to allow for large movement. The joint consists of alternate sheets of steel and rubber.

Fig. 60. Freyssinet flat jack.

should losses due to shrinkage render this necessary. The most recent design as used on a road at Winthorpe[4] in Nottinghamshire uses one jack which is inserted in a jacking pocket forming part of a prefabricated joint assembly which includes dowel bars as load transfer devices. After jacking, the slabs are retained in their stressed condition by means of grouting and the jacks can then be removed. Their place is taken by filling material or by grout which can be removed at any time for re-insertion of the jack if and when restressing is required. These jacking points have been provided at 500 ft intervals. Figure 61 shows the layout of one of these so-called "active joints" and Fig. 62 illustrates the jacking of the joint to impart prestress into the concrete. Some initial prestress is applied to prevent shrinkage cracking as soon as the new concrete is strong enough—often within 24 hours.

On an experimental road south of Paris at Fontenay-Tresigny[5] a number of other methods of applying prestress in slabs laid between abutments was used. One of these was the inclusion of a wedge-shaped

Fig. 61. "Active joint" assembly for prestressed concrete road jacked against abutments. The dowel bars are present to maintain a smooth running surface and to reduce the possibility of buckling. The concrete blocks are inserted to prevent closure of the joint when the jacks are removed.

Fig. 62. Jacks being inflated to prestress concrete road slabs at Winthorpe, Nottinghamshire. On completion of jacking, the joint is blocked and the jacks removed.

## 112 Concrete in Highway Engineering

piece of concrete across the road, a force to drive home the wedge being applied through prestressing cables acting against an abutment. This has also been used in Switzerland (Fig. 63). In another method the stress was applied through an inflatable rubber tube inserted in a joint and in a third a well-lubricated steel wire rope laid in a snake-like path across the road exerted a compressive force in the concrete when the rope was tensioned (Fig. 64). There is clearly much scope for inventive thinking in this field. As has already been pointed out, the great advantage of these and similar methods of applying prestress is that no

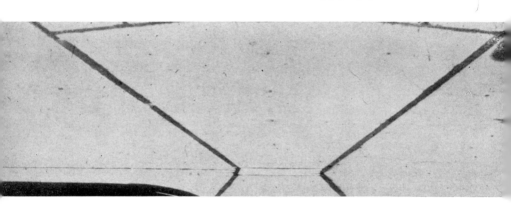

FIG. 63. Double wedge system for prestressing road slab in Switzerland. The wedges are drawn together by prestressing cables and the small central area of concrete is placed later.

moving joints are necessary. The use of concrete made with expanding cement has already been mentioned, but difficulties arise in controlling the expansion and in applying stress uniformly along the length. Figure 65 shows an experimental pneumatic joint which is designed to keep the amount of prestress constant, so eliminating the danger of buckling failure.

**Longitudinal Prestress Requirements**

With a subgrade restraint coefficient of about 1·0, the stresses in a concrete slab about 400 ft long will be of the order of 200–300 lb/in$^2$.

To allow for creep and any inequalities of prestress along the length it has been suggested that a minimum stress of 300 lb/in² is sufficient for a 400 ft slab. Stresses due to loads may be as high as 400–500 lb/in² at edges and corners and stresses due to warping may reach a maximum value of about 400 lb/in². Assuming that all three factors act together, tensile stresses of the order of 1000 lb/in² may occur in the concrete in the worst conditions. Although a number of existing prestressed roads have a much lower prestress than 1000 lb/in² it is now widely considered that a higher amount is desirable, particularly as it costs very little to obtain.

For roads of the continuous type with an initial prestress of 1000 lb/in² it is possible for the compressive stress to rise to 2500 lb/in² for a temperature range of about $-10°C$ to $40°C$, not unreasonable figures to design for in this country.

In practice, some buckling failures have occurred in prestressed slabs constructed on the continuous system. These failures occur at points where constructed defects such as poorly compacted concrete

Fig. 64. Experimental method of stressing in use in France. The snake-like duct contains a greased cable which is tensioned, thus exerting strain on the slabs.

Fig. 65. Experimental compressible pneumatic joint in continuous prestressed slab design.

Fig. 64.      Fig. 65.

FIG. 66.     FIG. 67.

FIG. 66. Inflation of Freyssinet flat jacks to re-establish prestress
in a runway at Maison Blanche.
FIG. 67. Freyssinet flat jacks in active joint, and transverse cables
in position ready for stressing.

are present or where detail design has been poor. Compressive stresses as high as 2500 lb/in$^2$ have been measured in practice in hot weather and it is clearly advantageous to use a limestone coarse aggregate in the concrete as this would considerably reduce the stress. Reduction of stress is not only advisable from the point of view of avoiding compression failure but also to reduce stress relaxation, due to creep. Experience of loss of prestress has been variable. In some cases losses have been high in the first two or three years, after which static conditions have been approached. In other cases, annual restressing has proved necessary to ensure an adequate amount of prestress is present in the coldest weather. Figure 66 shows the process of reapplication of stress through Freyssinet flat jacks left in the slab for the purpose at Maison Blanche Airport. The crack opened up over the jacks is subsequently sealed.

**Transverse Prestress**

Most prestressed pavements so far designed have included transverse prestressing (Fig. 67) as well as longitudinal stressing. The prestress has been induced by tendons and the amount has not usually exceeded 200 lb/in$^2$, often being very considerably less. It has been suggested that transverse prestress is not necessary and that fabric reinforcement would be sufficient or indeed no reinforcement at all provided a tied longitudinal joint is incorporated. Experiments in France and in Britain[6] are under way to investigate this point. Transverse prestressing accounts for a large proportion of the cost of the prestressing operation and important economies will result if it can be avoided.

**Costs**

Because of the relatively short length of prestressed concrete in roads so far built, an accurate estimate of costs cannot be formed. A number of engineers have, however, asked contractors to price prestressed designs and equivalent pavements designed by orthodox methods, and it seems that there is very little difference, at least as far as main roads are concerned. In Germany and in France prestressed concrete airfields have proved cheaper than normal concrete or flexible construction—a result which is understandable because of the considerable reduction in slab thickness that prestressing allows. Experience of prestressed pavements so far built seems to indicate that no special problems exist,[6] and when contractors become experienced in the stressing techniques and when more knowledge is available about thicknesses and amounts of prestress required, then costs will be comparable and perhaps even less than standard pavements. There exists the possibility that the performance of prestressed pavements will be considerably better for a longer period than orthodox pavements. If this proves to be true, this factor should be taken into account when assessing relative costs. The present state of development of prestressed roads is not encouraging, but there may be more scope for use in thick pavements as on airfields.

**References**

1. STOTT, J. P. Prestressed concrete roads, *Proceedings of the Institution of Civil Engineers*, Road Paper No. 47, Part II, October 1955.
2. PELTIER, R. *Contribution to the study of prestressed concrete roads*, Library Translation No. 102, London, Cement and Concrete Association, December 1962, pp. 35.
3. OBERTOP, D. H. F. Proefrak van 1000 m lang met voorgespannon betonplaten in Rijksweg 4A; proewen en metingen, *Cement*, June 1962, pp. 361–362.

4. LOCK, E. E. and STOTT, J. P. *The construction of an experimental prestressed concrete road at Winthorpe, Notts.*, Road Research Laboratory, Laboratory Note LN/593, June 1962, pp. 12.
5. PELTIER, R. *The tests on the prestressed concrete road at Fontenay-Tresigny*, London, Cement and Concrete Association, Library Translation No. 97, September 1962, pp. 41.
6. KIDD, R. A. and STOTT, J. P. The construction of an experimental prestressed concrete road at Winthorpe, Notts., *Proceedings of the Institution of Civil Engineers*, vol. 36, March 1967, pp. 473–498.

CHAPTER 7

# Maintenance and Repair Techniques

## Introduction

Some maintenance is required on all types of road surface. Repairs to roads may arise either because some part of the structure has failed or because an opening has been deliberately made into it—for example, to repair a defective service pipe. The repair techniques used for cement stabilised and concrete roads differ mainly because of the difference between the permanency of the surface material and they will, therefore, be considered separately.

## Cement Stabilised Bases

MAINTENANCE

No routine maintenance is needed for cement stabilised bases which will normally be virtually everlasting. The replacement of the superimposed surfacing material or an addition to it at intervals will be the only maintenance commitment which will arise with this type of road.

REPAIR OF BASE DEFECTS

Under abnormal circumstances, defects may arise in a cement stabilised base which may necessitate its partial or complete renewal over small or large areas. The removal of the defective material is simple although it will not normally be desirable or practicable to replace less than 4 to 6 in. of compacted thickness of stabilised material. Any attempt to add new cement stabilised material in thin layers 1 or 2 in. thick is inviting failure. The amount of bond developed between new and old work is very small and the thickness of any new material must be sufficient to provide the necessary stability through its own bulk.

It is important that all defective and weak material should be removed and that vertical faces should be formed at the junction of the old and new cement stabilised material. From consideration of elastic properties there is some theoretical advantage in using the same material and cement content for the replacement as was used for the original base, although some engineers favour a somewhat greater cement content to obtain higher strength and stability. Although this may not be strictly necessary, experience has proved it to be desirable, because in practice small-scale repair work is rarely carried out as well as the original large-scale project.

Particular care should be taken to obtain a high state of compaction in the new material. Vibrating or smooth wheeled rollers can be used for the mass of the work but around the edges, particularly adjoining the original pavement, special care should be taken and hand or mechanical trench rammers or vibrating plate compactors should be used.

The normal requirements for curing and protection in frosty weather apply, although there will be a very natural tendency to cut these to a minimum in order to open the road to traffic. If necessary, the new work can be surfaced immediately, that is, the same day as it is laid, and provided the thickness of surfacing is sufficient, no harm to the new work in the base will result. If this procedure is not carried out, normal curing methods should be adopted and traffic kept off for at least 3 days and longer in cold weather.

TRENCH REINSTATEMENT

No difficulties arise when trenches have to be opened in cement stabilised materials. The edges should be cut as vertical as possible and any old material which has been weakened or damaged should be removed. It is preferable to use for replacement a stabilised material made of the same aggregate and with the same or rather greater cement content.

The most important part of any trench reinstatement work is to make sure that the soil backfilling is well compacted to a high density and preferably to the same density and moisture content as the undisturbed surrounding material. This is not always easy but one or two elementary precautions can improve the result. For example, in order to reduce the possibility of the excavated material drying out or becoming wet, the work should be executed as quickly as possible so that the excavated material is exposed for the shortest possible time. Covering the excavated material is justified in some cases. Material should be placed in the trench in thin layers about 6 in. thick and well compacted by mechanical rammers. If the excavated material is unsuitable for backfilling, it should be replaced by new material of similar characteristics or by a granular material or by a cement stabilised material. The use of the latter, although relatively expensive, is often justified because of the complete immunity from settlement which results.

The surfacing material can be placed immediately upon the compacted base and this is probably the best procedure, after which the road can be opened to traffic at once.

**Concrete Roads**

Concrete roads that are properly designed and well constructed require little maintenance and will provide trouble-free service for a considerable number of years. Inevitably, however, even the best of concrete roads will eventually begin to show defects, but these, if taken in hand early, can be remedied, so prolonging the life of the road. For many years, it has been widely considered that it was very difficult or even impossible to do very much about repairing defective concrete roads. In the last decade, however, there have been significant developments in remedial measures and new tools and new techniques have been evolved so that the repair of various types of defects is both a practical and an economical proposition. Maintenance is confined to joint and crack sealing, whilst repair measures include the following:

1. Restoration of levels in slabs which have suffered settlement.
2. Replacement of failed slabs.
3. Overslabbing.
4. Surfacing with a bituminous carpet.
5. Patching of surfaces.
6. Repairing joints.
7. Correction of slipperiness.

Trench reinstatement is strictly speaking neither maintenance nor repair, but it is convenient to consider it in this chapter.

120  *Concrete in Highway Engineering*

The items listed above will now be considered in greater detail:

1. *Joint sealing.* This work, although small in amount and cost, should not be neglected because planned maintenance keeps cost at a minimum. To neglect maintenance is to invite progressive damage due to percolation of water though unsealed joints and to encourage ingress of foreign matter into the joints which can result in spalling at the edges during expansion of the concrete.

Replacement of the joint filler of expansion joints is seldom required. The top of the filler will usually be at a sufficient depth (about 1 in.) below the surface of the pavement to form a slot deep enough to hold new sealing material. Maintenance of an expansion joint involves removal of the old joint sealing material and its replacement with new. This work can be done most advantageously by using a plough or rotating disc to remove lumps of old compound, followed by a rotating wire brush to clean the side walls of the joint to ensure a good bond with the new sealing material. When the concrete is dry and the joints have been thoroughly cleaned, compressed air should be used to blow out dust before sealing. In cool weather the new sealing material should be left slightly below the surface, but in warm weather full sealing can be carried out. It is preferable to seal in two stages rather than to fill the sealing groove in one pass. Overfilling should be avoided because not only is it wasteful, but it causes an unsightly black stripe across the pavement. In addition, vehicle tyres readily pick up the excess material and distribute it widely.

Contraction joints and longitudinal joints are cleaned and refilled in the same way as expansion joints, although the width of the groove will usually be less. Some old longitudinal joints were simple shallow grooves formed by arrissing, and it is very difficult to maintain a seal over them. In cases such as this, rotary cutters may be used to open a deeper groove which will hold the sealing compound more satisfactorily.

2. *Crack sealing.* When resealing of joints is being planned, an inspection of the surface for cracks can conveniently be carried out. Hair cracks do not need attention, particularly in reinforced slabs, as they do not admit water. Wider cracks up to $\frac{1}{8}$ in. wide should be sealed with a free-flowing bituminous material, preferably at the same

time as joints are maintained or with a suitable resin. Before filling, all cracks should be cleaned out by blowing with a jet of compressed air. Wider cracks should in addition be cleaned, using a power-driven rotating wire brush and in extreme cases a rotary steel or carborundum disc. The object of widening the crack in this way is to facilitate the admission of sealing compound. Like joint sealing, work on cracks should be undertaken in the autumn in dry weather. Overfilling causes unsightly disfigurement of the surface and serves no useful purpose.

REPAIR TECHNIQUES

1. *Restoration of slab levels.* Settlement of slabs destroys the alignment and the riding quality and can result in objectionable steps at joints and cracks, particularly if load transfer devices are not included in the design. Slabs can be restored to their original level by raising them mechanically and by injecting cement grout or a bituminous mixture under pressure, so raising and undersealing in the one operation. In America and in this country, slabs are normally raised by pumping in grout under pressure (see Fig. 68), but in Germany slabs have also been raised by blowing in compressed air to support them pending the injection of grout.

Both processes involve the drilling of holes about 2 in. in diameter through the concrete slabs through which pipes are inserted. The spacing of these holes varies with the nature of the material under the slabs, but is generally on a 4–8 ft grid. The air pressure needed to lift slabs varies,

FIG. 68. Lifting of concrete slabs on a motorway. Cement grout is injected into holes drilled through the slab, the process being done in stages.

122     Concrete in Highway Engineering

but is generally of the order of 50 lb/in$^2$. Smaller pressures are enough to inject grout under the slabs. Neat cement, cement mortar, bituminous mixtures, dry sand and mud slurry have all been used in under-sealing, although experiments carried out by the Road Research Laboratory[1] have shown that subsequent movement of slabs was reduced more by the use of cement grout than by the use of bitumen and that this treatment gave a long life.

It is sometimes necessary to carry out the lifting in stages, allowing the grout to harden between operations. This procedure is necessary in order to prevent excess escape from edges, joints and cracks. To reduce as far as possible any jamming of adjacent slabs, work of this kind should preferably be done in the autumn when the subgrade is dry and the slabs are not fully expanded. Immediately after lifting and grouting, joints should be raked out and resealed.

2. *Replacement of failed slabs.* When a slab is badly cracked it is cheaper and more satisfactory to replace the whole or a part. The cause of the damage should be ascertained and dealt with before the repairs are carried out if it is likely to affect the new work. After breaking out the defective concrete, the subgrade and base should be examined. Any soft areas should be removed and replaced with suitable hard material.

When the slab has been removed, any dowel bars and tie bars revealed in the adjoining concrete should be straightened, cleaned and, in the case of the former, coated with a thin layer of bitumen and provided with a cap as for new work. If dowel bars have not been provided originally they can be inserted in holes in the old slab made by a rock drill or in a slot made by sawing and breaking out. After cleaning, the holes should be washed out with water and filled with 3 : 1 cement mortar or with epoxy resin mortar. The dowel bars can then be pushed in. The free end is then coated with bitumen and provided with a cap as in new work. New concrete is placed, including reinforcement if used, choosing a rather heavier weight than in the original work. If the original slabs were unreinforced, consideration should be given to the use of reinforcement in the new work. When new concrete is placed, care should be taken to ensure that no bridge of concrete extends across expansion joints either under or at the ends of the filler material.

If only part of a slab is renewed, the old concrete should be sawn

*Maintenance and Repair Techniques* 123

with a vertical face for at least 1 in. and the old surface carefully cleaned so that it is free of traces of dust. Any reinforcement projecting from the old work should be carefully preserved for a length of about 2 ft and incorporated in the new concrete.

In order to speed the work and keep traffic disruption to a minimum, a system has been developed in Germany of replacing defective slabs of one lane width by precast prestressed slabs. The slabs are made in a factory and transported to the site, where they are lifted into position as shown in Fig. 69. Provision is made for including dowel bars between the old and new slabs.

3. *Surfacing with a bituminous carpet.* When a concrete road reaches the end of its useful life as a running surface it can be covered with a bituminous carpet and given a new lease of life for an indefinite period. It should be noted, however, that there is little purpose served by applying a bituminous surfacing over slabs which are structurally unsound

FIG. 69. A precast prestressed concrete slab about 40 ft × 11 ft being lifted into position on a German motorway to replace a 25-year-old defective section.

and, in particular, on those which are rocking or which are badly cracked. The concrete must be sufficiently robust to act as a foundation, as a surfacing will contribute little strength to the slab.

Badly cracked and rocking areas can be dealt with either by injecting grout or by their removal and renewal using a similar type of concrete (or alternatively a lean concrete) carefully compacted into position. Where the slab is renewed any defects in base and subgrade should be dealt with as suggested in sub-section 2 above.

In a design where joints are very frequent it is unlikely that difficulties will arise due to reflective cracking appearing in the surfacing over these joints unless differential vertical movement is taking place. Where expansion joints are placed more than some 40 ft apart, however, the movement which occurs at them due to temperature changes is likely to cause cracking even under substantial thicknesses of bituminous material, and it is probably better in cases such as this to rake out the joint filler and sealing material and to replace it with a cement grout, thus blocking the road. This operation is preferably carried out when the slabs are at a low temperature to reduce the possibility of opening at these joints and care must be taken to fill the joint completely.

Attempts to reduce the amount of reflective cracking in bituminous surfacing by reinforcing the surfacing with steel fabric, by encouraging sliding of the surfacing on the concrete adjacent to the joints or by providing specially elastic surfacing material over and adjacent to the joints have proved helpful. They appear to delay the onset of reflective cracking but do not prevent it entirely.

4. *Overslabbing.* As an alternative to surfacing with a bituminous carpet, particularly when they are in very poor condition or where traffic is heavy, old concrete roads can be overslabbed with new concrete, so taking advantage of their appreciable residual value. Even when in a badly deteriorated condition the slabs retain a degree of rigidity and ability to distribute loads which should not lightly be abandoned as a capital asset. The new surfacing in concrete can be appreciably thinner than the old slab and it is preferable to separate the two rather than to attempt to bond them together. Work of this type has been carried out in many places, notably in Belgium, America and in this country.[2, 3]

No exact methods exist for designing the thickness of overslabbing concrete, but experience has shown that it should be between 4 in. and 8 in. depending upon the density of traffic. Before laying the new concrete, any slabs which have failed very badly structurally should be renewed and rocking slabs stabilised. Any defective areas of subgrade soil or drainage should be remedied so that the future of the new work is not jeopardised. In less serious cases, surface levels should be restored if necessary with a thin course of bituminous material and the new slabs laid on top of this.

One advantage of the separation of the new slabs from the old is that, if desirable, a new joint spacing layout can be adopted and this should be in accordance with normal design procedure. Dowel bars should not be put in joints when the slabs are less than 6 in. thick, because of the risk of cracking over them, but they are in any event not very necessary in this type of design because of the very good load transfer provided by the old slabs.[4]

5. *Patching of surfaces.* Areas of surface spalling sometimes occur on concrete road surfaces because of frost action, excessive wear or faulty construction techniques. In the past these areas have usually been patched with bituminous materials which look unsightly and are rarely satisfactory. In recent years considerable advances have been made in the technique of using concrete for patching small areas.[5-7] Other materials, notably the epoxy resins, have good bonding properties to concrete and they have the advantage that they can be applied in very thin layers.[8-11]

Tests have indicated that the most important single factor contributing to success when patching is the condition of the old surface—its cleanliness, texture and strength or soundness. The surface must be cleaned and be free of weak material and this is best done using mechanical tools of various types (see Fig. 70). Sound concrete must sometimes be removed to provide the proper thickness of resurfacing and a $\frac{1}{2}$ in. patch in concrete is the minimum recommended at the present time. A saw cut should preferably be made to provide a straight line and a vertical edge where the old and new surfaces will meet. After scarifying, the surface should be swept dry and then washed and brushed thoroughly with water. In areas where the old surface is sound and where scarifying

Fig. 70.       Fig. 71.

Fig. 70. Compressed air-driven hammers for removing the skin of defective concrete prior to resurfacing.
Fig. 71. A transverse crack in a prestressed concrete road sealed with an epoxy resin. Note the neat and unobtrusive appearance.

is not necessary to correct levels, the surface may be cleaned with detergent and acid. The surface is first washed with water, and oil droppings are removed by using detergent which is scrubbed vigorously into the surface, preferably with power brushes. After washing, the surface is cleaned and etched with concentrated hydrochloric acid having a specific gravity of $1 \cdot 16$ applied at the rate of 1 gallon/15 yd$^2$. When effervescence has ceased, the concrete is hosed down and surface water brushed off. This acid treatment is not essential but is desirable if there is a lapse of more than a few days between scarifying the surface and resurfacing with concrete. A 1 : 1 sand/cement grout is then brushed into the surface, leaving a film, and this is followed immediately with a concrete usually made with a maximum size of aggregate of $\frac{3}{8}$ in. The concrete is thoroughly compacted, preferably with a power-driven vibrating beam, the surface being finished by brushing. The use of an air entraining agent for this work is strongly recommended to give frost resistance. Any joints in the underlying concrete must be accurately reproduced in the surfacing. Curing by a resin based curing membrane is the most convenient practice.

Patching with concrete is cheap and effective where the larger areas are involved, but for thin layers over small areas, epoxy resin sand mixtures are very convenient, particularly as traffic can use the road within a few hours. The mixtures are tough, weather resistant and adhere well to concrete. In use, sand and the basic epoxy resin are mixed with a hardening agent which is formulated by the manufacturers to give the correct rate of setting for the purpose in hand.[8] The mixture is spread and worked over the areas to be patched, which are cleaned and prepared as already described. Although some success has been achieved with resin mixtures feathered out at the edges, filling up to a saw cut is clearly going to be more successful. Polyester resins and other formulations have also been used with success.

6. *Repairs at joints.* Defective joints can be rebuilt using concrete or epoxy resin mixtures, the choice depending on the volume and shape of the material to be replaced. It is in all cases essential to remove all defective or weak concrete so that the new material has a chance of being securely bonded and held to the old. All reinforcement, dowels, tie bars, old joint material and sealing grooves should be faithfully reproduced in the new material, which should be allowed to age for an adequate time to develop strength before the road is opened to traffic.

Joint repair work requires careful workmanship and a knowledge of the functioning of the various types and elements comprising joints.

FIG. 72. Defective concrete removed over a crack and new concrete being placed in position. Note the malleable sealing strip positioned over the crack to allow for future movement.

FIG. 73. The repair in Fig. 72 being completed.

FIG. 72.    FIG. 73.

When well carried out, the result can be equal in effectiveness and service life to the undamaged pavement and much of the deterioration seen in practice on roads in service could and should be remedied at an early stage.

FIG. 74. A pneumatic tool for breaking out defective concrete prior to repair.

7. *Repair at cracks.* The techniques for this type of work have now been extensively developed.[9, 10] If cracks are narrow and are unlikely to vary in width they can be sealed by a cement mortar, or by epoxy resin (Fig. 71). If movement is likely, a technique as illustrated in Figs. 72 and 73 is desirable. Figure 72 shows a longitudinal crack widened out and defective adjacent concrete being renewed. A malleable sealing strip is laid over the crack and a fine concrete filling well rammed in to pro-

## Maintenance and Repair Techniques 129

vide a new surface as shown in Fig. 73. Figure 74 shows the pneumatic tool used for breaking out the old concrete.

8. *Correction of slipperiness.* It is possible to improve the skidding resistance of smooth concrete surfaces by several methods. These include scarifying the surface by mechanical tools, etching the surface using concentrated hydrochloric acid, and roughening the surface using steel shot fed beneath a bulldozer blade. This last method is a cheap one as the shot is collected and re-used, but it is not effective for long periods under heavy traffic. Acid etching does not produce a permanent effect as the depth of treatment is small.

A specially developed piece of apparatus consists of a number of mechanically driven hammers attached to a chassis on wheels. The machine is moved slowly over the surface and removes the top skin of concrete, exposing a fresh surface. Another more recent machine is a power-driven drum studded with industrial diamonds which grind and pattern the surface. Grooving of slippery concrete surfaces has proved effective and power-driven machines have been designed specially for the purpose. The grooves should be formed at right angles to the passage of traffic to obtain maximum skid resistance.

Instead of renewing the concrete surface as described above, a new face of skid-resistant stone chips can be applied by surface dressing using bituminous binders or epoxy resins. A double surface dressing using a small chip for the first coat is recommended when tar or bitumen is used and good results have been obtained with rubberised bitumen binders and precoated chips. Surface dressing is not suitable for heavily trafficked roads.

TRENCH REINSTATEMENT

When reinstating a concrete slab after a trench opening, the same procedure for backfilling should be followed as is given for cement stabilised bases. It cannot be emphasised too strongly that the success of a reinstatement depends largely upon thorough compaction of the backfilling. The base under the slab should be reinstated using the same type of material laid and thoroughly well compacted to a similar thickness as in the rest of the road.

When the old concrete is being removed, it should be cut back to leave a shoulder of solid ground about 12 in. wide on each side of the excavation, leaving about 12 in. of reinforcing fabric projecting (where it is provided). The edges of the old concrete should preferably be sawn for the top 2 in. and then sloped towards the opening so that the new concrete will gain support from the old. Before placing the new concrete, all loose material should be cleaned away from the old concrete and the exposed surfaces brushed with a wire brush and moistened with water. The new concrete should be finished to a level slightly proud of the existing surface. A second tamping should be given an hour or two later to take up any shrinkage away from the existing concrete, the final surface being finished flush. The replacement concrete should always be reinforced with two layers of fabric, a layer being provided about $1\frac{1}{2}$ in. from the top and bottom surfaces. The fabric should span the opening with the heavier bars parallel to the line of the road. The weight of reinforcement in each layer should be at least equal to the weight of fabric originally provided.

WIDENING OF EXISTING CONCRETE ROADS

New slabs laid alongside the concrete roads should be tied with tie bars to the old work. Joints in the new work should coincide with those in the old. Care should be taken to provide an adequate base for the new work and the width should not normally be less than 4 ft. A sealing groove should be formed between the old and new work.

### References

1. KIRKHAM, R. H. H. Trends in research at the Road Research Laboratory, *Structural Concrete*, vol. 1, No. 6, November/December 1962, pp. 257–267.
2. DUTRON, P. and HUBERTY, J. *Resurfacing and widening old concrete carriageways with concrete*, Library Translation No. 88, London, Cement and Concrete Association, April 1960, pp. 11.
3. Reconstruction of Fosse Way in Leicestershire, *Roads and Road Construction*, vol. 36, February 1958, pp. 39–41.
4. PULLAR-STRECKER, P. J. E. H. *Crack and condition surveys of a section of trunk road A.46 in Leicestershire overslabbed in reinforced concrete*, Technical Report TRA 357, London, Cement and Concrete Association, November 1961, pp. 17.
5. WILLIAMS, R. I. T. *Thin bonded concrete surfacing applied to existing concrete road slabs*, Technical Report TRA 317, London, Cement and Concrete Association, April 1959, pp. 9.
6. FELT, E. J. The repair of concrete pavements, *Journal of the American Institute*, vol. 32, No. 2, August 1960, pp. 139–153.

7. KIRKHAM, R. H. H. and HIGGINS, G. E. Use of concrete for patching spalled areas in concrete roads, *The Surveyor and Municipal and County Engineer*, vol. 118, No. 3486, February 1959, pp. 151–154.
8. Guide for use of epoxy compounds for concrete, Report by ACI Committee 403, *ACI Journal*, September 1963.
9. LEIGH, J. V. Concrete carriageway repairs M1/M10 motorway—techniques used in three-year programme, *Highways and Public Works*, vol. 34, No. 1679, November 1966, pp. 8–10.
10. LEIGH, J. V. The motorway—some years later, *The Journal of the Institution of Municipal Engineers*, vol. 93, November 1966, pp. 358–363.
11. LEIGH, J. V. *Servicing a concrete motorway*, The Concrete Society Technical Paper PCS 17, September 1967.

CHAPTER 8

# The Appearance and Surface Characteristics of Concrete

**Introduction**

Concrete is the most widely used structural material we possess and because of its adaptability and economy, it follows that much of it will be exposed to view in various forms. It also follows that good appearance must be included as one of the essential qualities in addition to strength and surface characteristics. Considerable thought and effort must be expended in varying the surface finish and general appearance of the material, which is in itself functional and rather unsympathetic, particularly in rural surroundings. Not only must the appearance of the new concrete be considered but also the effects of age and weather. Much effort has been expended, particularly in the last 20 years, in devising surface treatments for concrete for the many situations in which it is exposed and some of these will be discussed in this chapter under the two headings of roads and structures.

**Road Surfaces**

Concrete has the unique quality of providing both the structural means of support and the running surface for road traffic. Fortunately the properties of concrete are such that the sound fulfilment of this dual purpose can readily be realised. In connection with the surface, the highway engineer and road user are interested in the following factors:

1. Riding quality.
2. Skid resistance.
3. Light reflection.

# The Appearance and Surface Characteristics of Concrete    133

4. Appearance and colour.
5. Icing tendency.
6. Ease of repair.

Points 1 and 6 are dealt with in Chapters 4 and 7 respectively and this chapter deals with 2–5.

## SKID RESISTANCE

The old conception of a skid-resistant surface as being very rough and knobbly has happily died as a result of research work, at least for concrete. Unlike bituminous surfacings, the skid resistance of concrete depends more on the properties of the fine aggregate than of the coarse aggregate, because the percentage of exposed coarse aggregate even on a well-worn concrete surface is relatively small. The use of some fine

FIG. 75. A typical brushed texture with an average texture depth of about 0·30 in. for high-speed roads.

Fig. 76. A machine for steel brooming fresh concrete surfaces to produce the surface texture desirable for high skid resistance on high-speed roads.

aggregates, e.g. limestone, can result in slippery surfaces, but fortunately most British natural sands are satisfactory.

An advantage of concrete compared with other surfaces is that there is less variation between different road surfaces, both from the point of view of age and of composition of the concrete. It was at one time believed that a heavily indented surface must be provided on concrete if good skid resistance was to be achieved. Experience has shown this to be unnecessary and a deeply patterned surface is in fact undesirable because of its bumpiness and because it sets up vibrations in vehicles. Deeply tamped surfaces are only justified on very steep slopes, where slow-moving traffic may find them advantageous.

The texture exposed to tyres on a new concrete road is that formed by the method of final finishing. For some time in this country a brushed finish has been used formed by drawing the bristles of a broom transversely across the slab. For high-speed traffic, a texture depth of about $0 \cdot 030$ in. as measured by a sand patch test appears to be desirable initially to allow for wear and this is achieved by a steel-bristled brush. Figure 75 illustrates the texture and Fig. 76 shows a mechanical method. Less vigorous finishes can be used in residential areas where traffic

# The Appearance and Surface Characteristics of Concrete   135

speeds are lower and these are formed by use of a bass broom. The skidding resistance of all road surfaces deteriorates with wear and concrete is no exception. If the skidding resistance becomes unsatisfactory, transverse grooving of the surface can be carried out mechanically at prices which are quite economic. Transverse grooves are effective not only in providing a new texture depth but also in assisting drainage of water from the road. Methods of grooving the plastic concrete are being developed.

## LIGHT REFLECTION

A bright road surface is essential as a background against which unlighted objects on the carriageway of a lighted street can be seen as dark silhouettes. From the point of view of street lighting, therefore, concrete is an excellent surface. Light concrete surfaces can be lighted at about half the cost necessary to achieve the same lighting result on dark-surfaced roads. Some countries, e.g. Germany and Switzerland, specify a required intensity of illumination in accordance with

FIG. 77. A concrete carriageway showing excellent light reflecting properties under headlights. The hard shoulder is also in concrete and has a rough texture formed by the inclusion of large calcined flint aggregate.

FIG. 78. An exposed aggregate road surface after 10 years of light traffic. The aggregate is $\frac{3}{4}$–$\frac{3}{8}$ in. pink Leicestershire granite.

FIG. 77.

FIG. 78.

the colour of the road surface. It follows that cheaper installations can be provided and maintained to light concrete carriageways than for black bituminous surfaced roads.

The advantages of concrete surfaces are not confined to lighted streets for it is common experience that concrete surfaces appear to be more brightly lit by car headlamps than do bituminous surfacings. This means that obstructions are more readily thrown up in silhouette and greater driving competence and safety are generated. Figure 77 shows the excellent result obtained on a motorway where a concrete carriageway and a concrete hard shoulder with a rough surfacing containing calcined flint were employed. It is sometimes held that the light colour of concrete is a disadvantage to drivers in bright sunshine. Whilst this is true of newly completed concrete road surfaces which can produce glare, this brightness soon diminishes under the action of traffic.

APPEARANCE AND COLOUR

The colour and texture of concrete can readily be varied by pigmenting, including darkening, and by the introduction of special aggregates which are exposed at the surface. With these two combinations of colour and texture almost any desired effect can be obtained, a fact which is probably not used to full advantage by highway engineers.

Lay-bys, acceleration and deceleration lanes, parking areas, lane markings—all these can be differentiated from the main areas of the carriageway. As an extension of this theme, on a bituminous surfaced trunk road in Switzerland the junctions were constructed in concrete to warn drivers, particularly at night, that they were approaching areas of hazard. Where a highly decorative finish is required, the aggregate in the concrete can be exposed (Fig. 78). There are two main ways of obtaining exposed aggregate finish. The coarse aggregate used in the concrete can be exposed or aggregate can be added to the surface of the concrete when it is in the plastic state so that it remains exposed. The appearance of the surface depends almost entirely on the size and nature of the coarse aggregate chosen. For example, granites can be used to produce a red or green finish and a gravel will give a brown coloured surface.[1] If it is desired merely to tone down or darken the surface of the concrete, this can be achieved by incorporating 2–4% of ferric oxide by weight of cement in the mix during the mixing process. Proprietary

## The Appearance and Surface Characteristics of Concrete

materials are available which are specially prepared for the purpose. For coloured work, coloured cement can be obtained from the manufacturers and the use of a coloured aggregate to match the cement will assist in providing a uniform appearance. It is desirable to lay small trial areas of all special finishes to develop the technique and to ensure that the colour and finish are as desired.

ICING TENDENCY

Concrete being light in colour gives up its heat slowly when cooling at night and absorbs heat slowly when exposed to warm atmospheres. It follows that ice and frost will form more slowly on concrete than on black surfaces, but once formed, frozen surface films will tend to disperse more slowly on concrete as air temperatures rise. In danger areas—for example, on slopes and at junctions—it is becoming increasingly common to embody electric heating wires near the surface of the concrete, not so much to melt frozen coatings but to prevent them forming.

## Structural Concrete

INTRODUCTION

Concrete is still often regarded, possibly because of its cheapness and extensive use, as a cheap expedient to be thrown by unskilled labour into formwork which is put together in a haphazard way with haphazard materials. Those who practise this thesis reap benefit accordingly.

The architectural use of concrete has undergone considerable development in the last 20 years under the stimulus of the need to use concrete not only as a structural medium, but also as a facing and decorative material. A position which originally arose due to the need for rigid economy under post-war building programmes, and because of shortage of material which, generally speaking, did not affect concrete, has now changed so that concrete is now recognised for its own inherent benefits. Engineers have been slower than architects in taking advantage of modern developments in concrete surface finishes, probably because the mass of most engineering structures has sufficient impact by itself without the need for surface embellishment. The position is now changing with the great increase in building work and engineers are themselves seeking ways of making their structures more interesting.

138   *Concrete in Highway Engineering*

Many engineers and architects have of course succeeded in producing concrete of excellent appearance but, at the same time, others have failed without being able to find reasons for their failure. Until recently factors affecting the appearance of concrete have not been sufficiently understood for specifications and construction procedures to be directed specifically towards achieving high quality appearance. As a result of research done by the Cement and Concrete Association and others, the causes of various blemishes have been established and means of preventing them found.[2-4]

THE CONCRETE MIX AND PLACING

It may be a paradox to say that in this aspect of concreting, the mix itself is probably of secondary importance. For many years, engineers have concerned themselves in producing lean, dry mixes by employing close control, but these features, although desirable in themselves, do not necessarily produce the best and most economical results on the construction site. This particularly applies in heavily reinforced and in prestressed sections where there is little concrete to compact. It is often cheaper to use a richer mix in these circumstances because it is more workable and is easily and consistently compactable. The generation of high internal pressures during placings assists in the removal of air entrapped on the face of the concrete and this may be achieved by rates of placings as fast as possible consistent with effective compaction. A vertical rate of not less than 6 ft per hour is recommended.

APPEARANCE OF CONCRETE

The colour of a normal, untreated concrete surface is an uninteresting, cold grey, and it is dependent almost entirely on the colour of the cement used. In a large civil engineering structure this appearance is not necessarily detrimental and considerations of form and shape usually over-ride the limitation. For this reason in civil engineering works, coloured concrete is not normally used, but it can be obtained either by the use of coloured cement, by exposed aggregate finishes, or by painting. The maintenance of uniform colour will be assisted by the use of cement from one works and by the use of fine and, of lesser importance, coarse aggregate from the same source.

The fact that concrete is a man-made material and exists in two

## The Appearance and Surface Characteristics of Concrete

conditions—the plastic and the hardened states—provides opportunities of producing a wide variety of surface textures, depending on the way it is moulded in the plastic state or the treatment given to the surface in its hardened state. Engineers in the past have usually used a smooth finish to their concrete, but there has recently been a trend to be more adventurous and to use textured surfaces. As a general rule, it can be said that a shaped or textured surface, whether produced by the material used for the formwork or by subsequent after-treatment of the surface, is aesthetically more satisfactory than a smooth surface. The reason for this is that the shape and texture of the surface to some extent camouflage discolorations due to mould oil and changes in colour of the concrete due to variations in the nature and surface characteristics of the formwork or mix.

COLOUR VARIATION AND STAINING OF CONCRETE

The appearance of many concrete structures is ruined by unsightly stains on the surface. A common cause is the use of an unsuitable release agent or mould oil or too much of the material. Some of the release agents available on the market are unsuitable because they react with the concrete or penetrate differentially into the concrete and the formwork. Generally speaking, the thinner the film of release agent, the better the surface finish, but the film must be complete. Chemical agents or oil phased emulsions—that is, emulsions of water in oil—are the most satisfactory for general use, although where steel linings are used it is preferable to use a neat oil containing up to $2\%$ of surfactant.

Another cause of staining takes the form of a dark line between panels. The trouble can be due either to the escape of mortar or grout at these places leaving a porous surface or by the opening up of the joint due to shrinkage of the formwork. Air reaches the face of the concrete and permits evaporation of moisture to take place on the line of the joint. In either case this loss of moisture or uneven drying out will cause a local darkening of the concrete. Tooling does not remove defects of this type.

Another cause of discoloration is that brought about by combining different quality materials in the same formwork panels. Concrete cast against timber which has a high absorption will be dark in colour, whereas concrete cast against timber which has a low absorption will be lighter. Untreated plywoods and timber have absorbencies varying widely across their surfaces and therefore, to produce uniformity in colour, must be thoroughly impregnated before use. This may require many applications of mould oil. Old and new timber mixed together in formwork will produce light and dark areas because of differing absorbencies. Where uniformity of colour is important, and the need to avoid pocking on the surface is of secondary importance, impervious linings such as plastics, steel, oil tempered hardboards or ply with impregnated surface veneer should be used.

In other cases, the need to avoid pocking is all important and colour variations can be accepted; in these cases absorbent linings of paper or card should be used. These are very cheap and are used only once. An advantage in their use is that they can be left in contact with the face of the concrete for curing purposes after the structural part of the shuttering is stripped.

Where surfaces of very high quality are required which are devoid of pocking and are uniform in colour, linings of moderate absorbency should be used. Examples are laminated card or standard hardboard, but these linings must be used only once. Where absorbent linings are used, the water content of the concrete and the rate of placing should be carefully controlled to avoid, as far as possible, colour differences between lifts and within each lift.

Stains on concrete occur due to water carrying rust down the face from reinforcing bars left exposed and from wire ties left exposed on the face. Wire cut-offs left on the forms cause rust staining on the soffits of concrete structures and all these marks are difficult to remove and are therefore best avoided.

*IN SITU* CONCRETE FINISHES

Finishes in *in situ* concrete can be produced:[5]

1. Direct from the formwork.
2. By exposing the aggregate.
3. By applying a finish.

FINISHES DIRECT FROM THE FORMWORK

The finishes to be obtained direct from the formwork are board marked finishes, smooth finishes and those produced from special formwork linings. A wide variety of board marked patterns and textures is obtainable depending on the type and quality of timber and the way in which it is used. Sound, straight, soft woods, free from large, loose knots, are normally used for practically all kinds of formwork and the impression of the grain in the concrete will depend upon the species of the wood and the manner in which it is sawn or re-sawn—that is, by band saw, framed saw or circular saw. In all work of this type it is important that specifications are carefully and precisely written to give contractors full information of what is desired. It is always advantageous to cast one or two large sized trial panels to determine exactly the procedure to follow. Boards must be uniform in width and thickness so that there will be no offset at the joints unless these are desired to form interest in themselves. Samples of concrete cast against different types of timber are given in Figs. 79 and 80.

Where it is desired to obtain a smooth final finish to the surface of the concrete it is necessary either to line the formwork with sheets of plywood or fibrewood or to prefabricate the formwork using sheets of plywood attached direct to studdings.

Special formwork linings can be made of rubber, plastic, polystyrene or similar material (Figs. 81 and 82). Casting concrete against thermoplastic sheeting produces a concrete with a smooth, eggshell-like finish. This finish is, however, likely to craze and will pick up quantities of dirt in time, although this may not be a disadvantage in clean environments

FIG. 79. Concrete finish obtained by using 7 × 1¼ in. rough sawn Douglas fir timber shuttering.
FIG. 80. Concrete cast against Douglas fir timber with the edges of the boards chamfered to produce a projecting V-bead, the aggregate being exposed.

FIG. 79.    FIG. 80.

Fig. 81.                                          Fig. 82.

Fig. 81. A deeply indented concrete finish resulting from timber formwork being used to break up a large area of concrete on a bridge abutment.

Fig. 82. An example of a bush hammered bridge abutment with precast concrete panels of exposed blue Shap granite aggregate under the handrail. The side of the beam is formed from profiled formwork with a smooth edge strip at the bottom.

or if the pattern is only to be viewed from a distance. Unlike most other materials, plastics require no release agent unless the surface is very scratched after many uses. The use of absorptive boards can be very successful because of the high quality, uniform and pleasing texture that can be obtained.

EXPOSED AGGREGATE AND TOOLED FINISHES

The exposed aggregate finish reveals a truer nature of concrete as a material and can give varying degrees of colour and texture, depending upon the aggregate used. An exposed aggregate finish, however, will not be applicable to every type of structure, and it must be remembered that any imperfections in the concrete such as honeycombing, poorly made construction joints, loss of water between individual boards or panels will generally become more noticeable when the aggregate is exposed. In addition, if the formwork is made up in panels, the pattern of these panels will invariably be reproduced on the face of the concrete after it has been tooled, irrespective of whether or not there has been an escape of water between the panels. The simplest and cheapest exposed aggregate finish is obtained by stripping the formwork at an early age and washing and brushing the surface to expose the aggregate.

The period up to which this method is effective depends considerably on the time of year, but in temperate conditions, 16–18 hours will usually be found to be the upper limit.

After this period the matrix will have become too hard for removal. Steel bristle brushes are normally used for the work, which is started at the bottom, the work being carried upwards. The application of a fine spray of water during the brushing process is advantageous. Where formwork is to be left in position for periods of longer than 18 hours the set of the cement can be retarded by treating the formwork with a retarding agent. A number of proprietary materials are on the market, but their use needs care to obtain a uniform distribution of the retarder over the formwork. The softened material is removed by washing and brushing as before.

Concrete can be tooled to provide a variety of finishes. Bush hammering, point tooling and chiselling are methods in widespread use. Power tools which are either electrically or pneumatically operated are used for working on *in situ* concrete with any fine, detailed work being done

FIG. 83. Point tooling on a bridge beam and pier in contrast to the rough texture of the sandstone block facing and the smooth edges to the beam.
FIG. 84. Bridge approach span piers with an aggregate transfer finish.

FIG. 83.

FIG. 84.

by hand. Aggregates which tool without shattering such as granites and other igneous rocks are more successful than natural gravels (Fig. 83). The colours and textures will to some extent be governed by the type of aggregate, but the more important factor is the type of tool used. Mechanical grinding of concrete surfaces is expensive and is not often used for engineering structures.

The aggregate transfer method is a means whereby an exposed aggregate finish can be given to the face of *in situ* concrete. As has been emphasised, the success of the methods already described depends on the use of first-class formwork, a high degree of control over the concrete mix and the placing and compaction of the concrete within the forms. By the aggregate transfer method difficulties arising from differentials in these requirements can to a very large extent be overcome. The process consists of sticking selected aggregates to sheets of hardboard which become form liners. A water-soluble cellulose adhesive is used and a uniform spread of chips is aimed at. Panel effects can be obtained if desired by sticking strips of timber to the face of the hardboard. The concrete is placed within the formwork in the usual way and after hardening the forms are removed, leaving the aggregate forming part of the face. The adhesive and sand mixture covering the face of the aggregate can then be removed by scrubbing and washing down with clean water[6] (see Fig. 84).

Sand, grit and shot blasting are sometimes used to expose the aggregate on the face of the concrete. Decorative effects can be obtained by varying the depth of blasting or by leaving some parts unblasted in regular or irregular patterns. This process has the advantage that the timing of the operation is not critical.

With all exposed aggregate finishes, particularly those where the surface of hardened cement paste has been removed by mechanical tooling, the cover to reinforcement should be increased to compensate for the material removed from the face of the concrete and the possible disturbance of the bond between the particles of aggregate near the face and the cement paste holding these particles in place.

APPLIED FINISHES

Applied finishes include rendering, proprietary cement, plastic, and aggregate mixtures and painting, but none of these is often used for

*The Appearance and Surface Characteristics of Concrete* 145

highway structures because of their expense. Painting has been used effectively in some cases sometimes to give a contrast in lightness, for example to the vertical side of an edge beam under a projecting coping. Cement paints, emulsion paints and chlorinated rubber based paints have all been successful.

PRECAST FINISHES

The use of precast finishes as opposed to treating *in situ* concrete has the great advantage that not only is the variety of finish which can be readily achieved much greater, but also any work which is not entirely satisfactory can be readily rejected. Many of the techniques developed for building purposes can be used by the highway engineer.[7, 8] The following list gives some idea of the types of finish that are available for highway use.

(1) *The use of white or coloured cements* in place of ordinary Portland cement with or without a special sand with the object of changing the colour of the concrete. It is entirely satisfactory to use concrete of strong uniform colour in large areas, but better results are often obtained where the face of the slab is heavily patterned or profiled so that any unevenness is masked by the patterns of the surface. Only bold, large patterns would normally be applicable for use in highway structures.

(2) *Casting concrete face down against special linings*, e.g. rough timber boarding, rubber, thermoplastic, etc., in order to impart a pattern or texture. A very large range of patterns and textures is clearly possible by this method.

Thermoplastic materials have the advantage that they require no release agent. As against this, however, the surface so produced is smooth and tends to craze much more readily than concrete cast against more absorbent materials. If the thermoplastic sheeting is deeply indented as shown in the example in Fig. 85 an appearance rather like a heavily tooled large stone block can be obtained which is very suitable for bridge abutments. Various forms of rubber matting have been used to impart pattern and texture to the face of concrete. Other methods include the use of timber which has been specially sawn—for example, across the grain in order to give a very rough surface—and screwing

Fig. 85. Fig. 86.

Fig. 85. A bridge abutment cast against thermoplastic formwork to produce a very rough finish akin to heavily tooled masonry.
Fig. 86. Precast concrete slabs faced with exposed grey Cornish granite aggregate used as facing to an underpass.

fillets of timber or laths to the shuttering. All these methods assist in reducing the scale of large masses of concrete and in providing additional surface interest.

(3) *Casting face up and producing a pattern or texture* to the face of the plastic or hardened concrete. This can be done by scraping the concrete, rolling it with an indented roller, by screeding or by the use of abrasive wheels or percussion tools. Other finishes can be produced by using a serated screeding board to produce lines either vertical or horizontal on the face of the slabs.

(4) *Treating the face of the plastic or hardened concrete* to expose the aggregate by washing, brushing, acid etching, tooling, grinding, sand blasting or shot blasting. Considerable development has gone into the production of exposed aggregates and because of the great variety of building and road stones fortunately available in this country, a wide range of colour and texture can be obtained. An additional advantage of these finishes is that, provided the correct aggregate is used for the particular situation, the finishes weather extremely well. Slabs can be produced either by using special facing mixes incorporating the coarse aggregate it is designed to expose, applied monolithically with the backing concrete, or by spreading a layer of stones over the concrete and tapping these into the surface. The aggregate is exposed by washing

# The Appearance and Surface Characteristics of Concrete

and brushing some 3 or more hours after casting, or the slabs are wire brushed by hand when the concrete is about 16 hours old or mechanically brushed when the concrete is older. Figures 86 and 87 show various uses of exposed aggregate finishes.

(5) *Facing the concrete with aggregate particles, tiles, mosaic or special material* such as glass and ceramic pieces applied to the hardened concrete or used as a lining to the mould. These finishes, although used for architectural finishes, are not widely applicable for highway use. They do, however, weather well.

## CHOICE OF SURFACE FINISH

Because many parts of Britain have atmospheres charged with sooty and tarry particles, the preservation of a good appearance is dependent upon the correct choice of facing slabs. For this reason, exposed aggregate slabs weather well because of their rough texture, which masks the deposit of dirt and grime on their surfaces. The more rounded particles of aggregate with a glassy texture such as river or seashore gravels are most successful in this connection. It is always necessary to pay special attention to the paths of rainwater flowing over the face of the concrete and the use of precast panels makes the detail problem easier. Emphasis is usually given to the edges of panels by the formation of margins in a contrasting texture. Joints between panels do not usually present any problems when the slabs are used as permanent shuttering.

FIG. 87. Precast concrete panels used to face a retaining wall.
FIG. 88. Split concrete blocks used as formwork and permanent facing to a bridge abutment. The blocks are hydraulically pressed and weather attractively.

FIG. 87.

FIG. 88.

Fig. 89. Precast concrete exposed aggregate panels used to face a dividing strip on a central reservation. The cobblestones are set in concrete.

The use of a local aggregate as opposed to one imported from another district creates a structure with some affinity to the landscape. Figure 88 shows the use of split concrete blocks to face a bridge abutment—another solution to the problem which is both effective and inexpensive. Figure 89 illustrates the choice of bold exposed aggregate texture in a structure where mud and water spray would hide a smoother finish.

HANDLING AND FIXING PRECAST SLABS

The edges and arrises of precast concrete slabs are always liable to damage and should be carefully protected against chipping and breakage. The slabs should be carefully set down after lifting, care being taken to ensure that hoisting cables do not contact the finished faces. Slabs should not be moved about with crowbars unless the concrete is adequately protected by pads. Suitable spreaders should always be used when handling slabs by mechanical lifting devices. Chamfered edges are invaluable for hiding irregularities in alignment and in preventing edge damage.

On site there should be ample space for stacking and storing precast concrete slabs. They should be arranged so as to be easily accessible when required for fixing. Slabs should never be walked upon, nor should they be brought into contact with the ground, handled with

## The Appearance and Surface Characteristics of Concrete 149

cables or ropes or handled by dirty or greasy hands. The finished surface should always be protected from other construction operations and from the weather by covering them with non-staining waterproof or plastic sheeting. Slabs which have been partially weathered because they have been stacked on the outside of a pile will have a different appearance from those which have been protected by being placed inside a stack.

In bridges and retaining walls precast slabs are often used as permanent shuttering. They should be supported by battens held by suitable framing whilst the concrete is being cast behind them. Care must be taken to grout up the joints to prevent the mortar from the backing concrete escaping and running down the face, with consequent disfigurement. Steel or non-ferrous metal ties cast into the slabs and then incorporated into the backing concrete provide adequate fixing. Casting the backs of the slabs so that they are rough or forming key grooves in them are additional precautions in ensuring a good bond between the slabs and the *in situ* concrete. Care must be taken not to spill concrete over the surface of the slabs when it is placed behind them. A number of special jointing and supporting methods have been developed for precast slabs, the choice depending upon the type of slab and the situation where they are being used.[8]

### References

1. CEMENT AND CONCRETE ASSOCIATION. *Specification for housing estate and other minor roads in concrete*, Series Df. 1, London, September 1961, pp. 29.
2. KINNEAR, R. G. *A classification of the surface defects and some particular influences of formwork linings, release agents and concrete pressure on the appearance of concrete finishes*, Technical Report TRA 380, London, Cement and Concrete Association, July 1964, pp. 36.
3. MURPHY, W. E. Some influences of concrete mix design and method of placing on the surface appearance of concrete, *Symposium on surface treatment of in situ concrete*, London, Cement and Concrete Association, September 1964, pp. 40.
4. BLAKE, L. S., KINNEAR, R. G. and MURPHY, W. E. Recent research into factors affecting the appearance of *in situ* concrete, *Symposium on surface treatment of in situ concrete*, London, Cement and Concrete Association, September 1964, pp. 21.
5. WILSON, J. G. *Exposed concrete finishes*. Vol. I, *Finishes to in situ concrete*, C.R. Books Ltd., London, 1962, pp. 144.
6. CEMENT AND CONCRETE ASSOCIATION. *Aggregate transfer*, Series Cz. 4, London, March 1960, pp. 12.

7. WILSON, J. G. Surface finishes to large precast concrete slabs, *Proceedings of the conference Housing from the factory*, London, Cement and Concrete Association, October 1962, pp. 151.
8. WILSON, J. G. *Exposed concrete finishes*. Vol. II, *Finishes to precast concrete*, C.R. Books Ltd., London, 1964, pp. 184.

CHAPTER 9

# Construction in Extreme Conditions—Hot and Cold Weather

CONCRETE pavements can be laid throughout the winter months and also in high ambient temperatures if certain precautions are taken. Because of their physical dimensions, which are essentially those of a large thin sheet, more care and precautions are necessary when pavements are constructed than will be the case for most other types of structure. In climates which experience very cold and warmer seasons, it is always worth considering whether the extra precautions necessary to enable construction to proceed in the winter are worth while, bearing in mind that with present-day construction methods very large areas of pavement can be constructed in quite short times. It is quite easy to construct, for example, all the concrete necessary on an 8-mile dual carriageway motorway in about 40 working days, and with this kind of output it is clearly not necessary to construct during cold weather.

**Hot Weather Concreting**

Concreting during hot weather requires good planning and certain precautions. These are needed for a number of reasons. High temperatures accelerate the hardening of concrete and more mixing water is generally required for the same workability. This in turn often means adding additional cement in order to maintain the same water/cement ratio and strength. Higher water contents also mean greater drying shrinkage. In very hot weather fresh concrete may be plastic for an hour or less before it hardens. By exercising good control, however, concreting can proceed smoothly during hot weather and the final result can be satisfactory. The optimum temperature of fresh concrete is probably about 55–60°F, but this cannot be obtained without artificial cooling in hot weather. A higher temperature can be tolerated and 90°F is a reasonable and practical upper limit. Difficulty, however, can be experienced even with concrete temperatures of less than 90°F in a combination of hot, dry weather and high winds.

### EFFECTS OF HIGH CONCRETE TEMPERATURE

If the temperature of fresh concrete is increased from 50 to 100°F, an additional 4 gallons of water is needed for each cube yard of concrete to maintain the same workability. Increasing the water content of concrete without increasing the cement content results in a higher water/cement ratio, thereby adversely affecting the strength and other properties of the hardened concrete.

Tests have shown that while higher concrete temperatures increase early strength, at later ages the reverse is true. As an example, the 28-day strength of a concrete cast and kept at 120°F will be only about 80% of the same concrete cast and kept at a temperature of about 70°F. On the other hand, the reduction in strength of a concrete prepared and kept at 90°F will only be a few per cent lower than a concrete which is made and kept at about 70°F. These differences do not appear to increase with time and, in fact, the reverse is probably true.

Besides reducing the strength and increasing the mixing water requirement, high temperatures in fresh concrete produce harmful effects. Setting is accelerated, the rate of concrete hardening is increased and the length of time within which concrete can be handled and finished is shortened. In hot weather the tendency for cracks to form is increased both before and after hardening. Rapid evaporation of water from hot concrete may cause plastic shrinkage cracks to appear before the surface has hardened. Cracks may also develop in hardened concrete because of shrinkage due to higher water requirement or volume change due to cooling from the initial high temperature. Control of the air content of air entrained concrete is somewhat more difficult at elevated concrete temperatures. More air entraining admixture is required to produce a given air content. In listing these disadvantages, however, it must be emphasised that the effects are usually marginal and in practice are not critical to the conduct or quality of the work.

## COOLING CONCRETE MATERIALS

Operations in hot weather should be directed towards keeping the concrete as cool as practicable. The most practical method of maintaining low concrete temperatures is to control the temperature of the concrete materials. Because of the comparative quantities of materials involved, concrete temperature is primarily dependent upon the aggregate temperature. Of the materials in concrete, however, water is the easiest to cool and is the most effective pound for pound for lowering the temperature of concrete. Mixing water from a cool source should be used and it should not be stored in tanks exposed to the direct rays of the sun. Tanks and pipelines should preferably be buried and protected in some way. Painting exposed tanks white helps to prevent similar absorption of solar heat. Water may be cooled by refrigeration or by adding ice, but these extensive measures will rarely be justified for normal work.

Aggregates have a pronounced effect on the fresh concrete temperature because they represent 60–80% of the total volume of concrete. There are several simple methods of keeping aggregates cool. Stockpiles can be shaded from the sun and kept moist by sprinkling, which is effective as a cooling process by evaporation, especially when the relative humidity is low. Cement temperature has only a minor effect on the temperature of the freshly mixed concrete because of the low specific heat of the material and relatively small amount of cement in the mix. Since cement loses heat slowly during storage it is often hot when delivered, the heat being produced by grinding of the cement clinker during manufacture. Since the temperature of the cement does affect the temperature of the fresh concrete to some extent, specifications sometimes place a limit on its temperature at time of use. It is, however, more realistic to specify a maximum temperature for freshly mixed concrete rather than to place a limit on the temperature of the ingredients.

## PREPARATION BEFORE CONCRETING

Before the concrete is placed any precautions that can easily be taken to help lower the concrete temperature are worth while. These include the use of white paint on equipment and shading of the operations. The sprinkling of the subgrade, the reinforcing steel and the forms is helpful in cooling the surrounding air and increasing its relative humidity. There should, however, be no standing water on the subgrade when the concrete is placed. In extremely hot periods in arid climates, improved results may be obtained by restricting concrete operations to the early morning, early evening or entirely to night-time.

TRANSPORT, PLACING AND FINISHING

Transporting, placing and compaction should be done as quickly as possible during hot weather. Delays contribute to loss of workability and increased concrete temperatures. Prolonged mixing or agitating should be avoided because this increases the temperature of the concrete by mechanical attrition. If delays occur the heat generated by mixing can be minimised by agitating intermittently.

The surface of the concrete will tend to lose water very rapidly in hot weather, which can lead to difficulties such as plastic cracking and early shrinkage. Temporary sunshades and windbrakes help to minimise these problems and in extreme conditions it may be necessary to use a fog spray on the surface of the concrete till hardening has occurred. Alternatively, the concrete can be covered with impervious material such as plastic sheeting immediately the surface has been completed, although the film must be supported clear of the surface until the concrete has hardened in order to prevent marking. When covers are used they should be carefully positioned so that air is not encouraged to pass over the surface, so increasing the evaporation rate.

CURING AND PROTECTION

Curing and protection are more critical in hot weather than in cooler periods. Exposed concrete surfaces should be prevented from drying by commencing curing operations as soon as practicable. Continuous moist curing is preferable during hot weather and it should continue for at least 24 hours by, for example, covering the concrete with hessian and spraying it with water. As an alternative, concrete can be covered with curing paper or plastic sheets, care being taken not to mark the surface. It is essential, however, to prevent the movement of air under the covering, as this encourages evaporation of moisture from the surface. If curing compounds are used white pigmented materials are advantageous as they reduce the amount of solar heat absorption.

## ADMIXTURES

Admixtures are sometimes used during hot weather to delay the setting time of concrete and lessen the need for an increase in mixing water. Water-reducing set retarders may be specially beneficial, but trials should be carried out beforehand in order to establish their effect, if any, on the strength development and other properties of the concrete.

## Cold Weather Concreting

Concrete can be placed throughout the winter months or in cold climates if certain precautions are taken. In its early life concrete can be damaged by rain, sleet and snow, but the most severe climatic hazards are freezing temperatures and strong winds. Freshly placed concrete is vulnerable to freezing temperatures before and after it has stiffened. Hence, there are two stages to consider:

1. If the concrete while plastic is allowed to freeze, expansion of the water as it freezes can cause severe damage which will make the concrete useless.
2. Concrete which has stiffened but not attained sufficient maturity can be permanently damaged by the pressures developed by the free water freezing, thereby weakening the bond between the aggregate particle and the cement paste. Ultimate strength of such concrete may be reduced by as much as 80%.

The damage to concrete frozen when plastic is generally obvious as it will be in a disrupted state, but the damage to concrete frozen after it has stiffened but while it is still immature may go undetected. It is, therefore, necessary to ensure that freshly placed concrete is maintained at a temperature above freezing, both during and after the plastic stage, until such time as it has developed sufficient strength to resist the disruptive forces associated with freezing temperatures.

Strong winds increase the vulnerability of concrete because the heat loss from the concrete increases rapidly as the wind speed increases. Additionally a strong wind blowing across an unprotected concrete surface will increase the rate of evaporation of water and in these conditions water may freeze at an air temperature of above $32°F$.

Unprotected concrete surfaces are also vulnerable in still air conditions if the sky is clear and there is no cloud cover, as in these conditions heat loss by radiation can be very rapid. Wind and cloud conditions, therefore, should also be considered in conjunction with air temperature when assessing the need for protective measures.

EFFECT OF LOW CONCRETE TEMPERATURES

Temperatures affect the rate at which hydration of cement occurs, low temperatures retarding concrete hardening and strength gain. For example, the 28-day compressive strength of a concrete mix cast and cured at 40°F is about 15% less than the same mix cast and cured at about 70°F. The strength at one year of the concrete made at the lower temperature, however, will be about 10% greater than the second mix mentioned above. The strength gain of concrete practically stops when temperatures are very low, i.e. below about 35°F. Concrete of low workability is particularly desirable for cold weather road work. During cold weather evaporation, rates are reduced and unless water contents are minimised delays in finishing will occur.

INITIAL TEMPERATURE OF CONCRETE

The compressive strength level appropriate at the completion of the required prehardening period is in the range 500–650 lb/in$^2$, and in practice a figure of 750 lb/in$^2$ is recommended. This figure applies to all concretes irrespective of the type of cement used and is not affected by the addition of accelerated admixtures. Concrete that achieves this strength will not be affected by freezing conditions.

When it is proposed to place concrete at or below an air temperature of 36°F on a falling thermometer, it is essential to know the minimum initial temperature of the concrete when placed that will ensure that the "maturity" of the concrete is at least equal to the required prehardening period before its temperature falls to freezing point. This initial temperature should never be less than 40°F. Alternatively, when the maximum available initial concrete temperature is known, a decision can be made on the amount of protection the hardening concrete will require to ensure its safety. Pink[1] has shown how these simple calculations and decisions can be rapidly made.

## HEATING CONCRETE MATERIALS

The temperature of cement and aggregates varies with the weather and type of storage. Since aggregates usually contain moisture, frozen lumps and ice are often present when the temperature is below freezing. Frozen aggregates must be thawed out to avoid settlement of aggregate in the concrete after placing and to facilitate proper batching. At temperatures above freezing it is seldom necessary to heat aggregates. At temperatures below freezing the fine aggregate only is often heated to produce concrete of the required temperature. If aggregate temperatures are above freezing, the desired concrete temperature can usually be obtained by heating only the mixing water.

Circulating steam through pipes over which aggregates are stockpiled is a recommended method of heating aggregates. The stockpiles should be covered with tarpaulins to retain and distribute heat and to prevent formation of ice. The average temperature of the aggregates should not exceed about 150°F.

Of the ingredients used for mixing concrete the mixing water is the easiest and most practical to heat. It is, moreover, the most effective as water can store about five times as much heat as can solid materials of the same weight. The water, however, should not be hotter than 180°F because of the possibility of causing a flash set of the concrete. Attempts should never be made to heat the cement, although hot cement as supplied by the cement maker does contribute to a raising of the concrete temperature. The cement, however, does not make a particularly valuable contribution because of its low specific heat and because of the relatively small quantity used in the mix. The concrete temperature should not exceed 80°F.

## PREPARATION BEFORE PLACING CONCRETE

Concrete should never be placed on a frozen subgrade since uneven settlement may occur when thawing takes place. When the subgrade is frozen for only a depth of a few inches, thawing may be artificially induced by applying steam or heat from other artificial sources. The inside of forms and reinforcing steel should be free of snow and ice at the time concrete is placed.

## CURING METHODS

Concrete covered with insulation seldom loses enough moisture to impair curing. However, moist curing is needed during winter to offset drying when heat is applied and it is important that concrete should be supplied with ample moisture when warm air is used. Steam exhausted into an enclosure is an excellent method of curing concrete because it provides both heat and moisture. Early curing with membrane compounds may be used on slabs within heated enclosures, but it is better practice to keep the concrete moist first and then apply the curing compound after the heating is removed. Heat may be retained in the concrete by use of insulating blankets such as straw or glass wool made into sheets with polythene. The effectiveness of insulation can be determined by placing the thermometer under the insulation in contact with the concrete. Corners and edges of slabs are most vulnerable to freezing and checks should be carried out at these positions. The manufacturers of insulating materials can provide advice as to the effectiveness of their materials. It is essential that insulating materials should have a moisture-proof covering to withstand handling and to prevent absorption of moisture. Wet insulating materials are useless.

After the concrete is in place it should be kept at favourable curing temperatures until it gains sufficient strength. To enable the development of strength to be assessed cubes can be made and kept alongside the slab. They should be crushed at intervals in order to establish the rate of gain of strength with time. Accelerators are not normally used in concrete road slabs except for small-scale work. Air entrained concrete is valuable in the context of cold weather work because it is less susceptible to damage by early freezing than concrete without entrained air. It should be remembered, however, that whilst air entrained concrete gives protection against cold weather in service, it does not provide a complete protection during the hardening and maturing period.

## Cemented Materials in Hot and Cold Weather

### HOT WEATHER

Many of the comments made concerning concreting in hot weather also apply to cemented materials. The time available for placing and compaction will be reduced compared with work carried out in normal temperatures because of the increased rate of the evaporation of water and so greater limitations on time for working may be necessary. Cemented material which has lost moisture will attain a low crushing strength and will be poorly compacted. The protection of the surface during the curing period clearly is of considerable importance if the hydration of the cement is to proceed normally.

COLD WEATHER

The laying of cemented materials in sub-zero temperatures is not normally a very practical proposition. Moreover, the rate of gain of strength of cemented materials is very much less than that for concrete and the "safe" crushing strength of 500 lb/in$^2$ already quoted would take considerable time to achieve.

**Reference**

1. PINK, ALAN. *Winter concreting*, London, Cement and Concrete Association, 1967, pp. 24.

# Index

Acid etching 126, 146
Admixtures 11, 155
Aggregate transfer 144
Air entrained concrete 11, 12, 49, 50, 126, 152, 158
All-in aggregate 8, 39, 97, 101
Alternate bay construction 53

Bituminous surfacing 36, 37, 102
BLAKE, L. S. 46
BROOK, K. M. 46
BROOKE-BRADLEY, H. E. 83
Bush hammering 143

Calcium chloride 3, 12
C.B.R. design method 24, 26
Cement
  coloured Portland 1, 136, 137, 138, 145
  extra rapid hardening Portland 3, 18, 86, 94
  high alumina 3
  hydrophobic 3
  low heat Portland 2
  masonry 3
  ordinary Portland 1, 48, 94
  Portland blast-furnace 2, 94
  rapid hardening Portland 1, 94
  sulphate-resisting Portland 2
  supersulphated 3
  temperature 14, 153
  white 1, 145
Coarse aggregate 4
Cold weather concreting 13, 16, 155
Compaction of concrete 9, 10, 50, 62
Concrete mix design 6, 7, 48, 49, 62
Continuous reinforced slabs 36
Cores 64, 71

Cracking 29, 103, 120, 124, 128, 152, 154
Curing 15, 18, 58, 74, 88, 100, 158

Design of pavements 21, 36
Drainage 38, 41, 42, 43
Durability 5, 6, 10, 50, 85, 86

Epoxy resin 126, 127, 128

Fine aggregate 4
Flash set 4
Forms 53, 65
Freezing 6, 118
Frost damage 11, 125
Frost susceptible soils 25, 26, 40, 41, 86

Grading of aggregates 7
Grit blasting 144, 146

Hot weather concreting 13, 16, 151

Impurities in aggregates 5

Joint filling compounds 74
Joints and joint spacing 29, 32, 36

Limestone aggregate 5, 63, 114, 134

Mould oils 139

# Index

No-fines concrete  44

Paver mixers  64
PINK, A.  156
Plastic cracking  15, 16
Polyester resin  127
Profilometers  76
Protection of concrete  15

Ready mixed concrete  51, 52
Reconstituted aggregates  8
Reinforcement  28, 29, 31, 69
Retarders  12, 13, 15, 143
Riding quality  6, 76, 79

Sands  4
Sawing joints  6, 57, 73, 125
Scraping straight edge  55, 79

Segregation  8, 94
Skidding resistance  5, 55, 129, 133, 134, 135
Slag aggregate  5, 39
Slip form paver  77, 78
Spalling of surface  125
Strength of concrete  10, 11
Sulphates in soil and ground water  2, 3, 4
Surcharge of concrete  53
Surface water drainage  43, 44

Tongue and groove joints  35
Traffic loading  23, 24, 27
Trial lengths  71

Water for concrete  6
WESTERGAARD  27
Wet conditions  6
Workability of concrete  6, 7

*OTHER TITLES IN THE SERIES IN CIVIL ENGINEERING*

Vol. 1 ARUTYUNYAN: Some Problems in the Theory of Creep in Concrete Stuctures
Vol. 2 WALLER: Building on Springs
Vol. 3 SACHS: Wind Forces in Engineering